Stephen. [from old catalog] Roper, Edwin R. Keller, Clayton W. Pike

Roper's catechism for steam engineers and electricians ..

Stephen. [from old catalog] Roper, Edwin R. Keller, Clayton W. Pike

Roper's catechism for steam engineers and electricians ..

ISBN/EAN: 9783743467019

Manufactured in Europe, USA, Canada, Australia, Japa

Cover: Foto ©berggeist007 / pixelio.de

Manufactured and distributed by brebook publishing software (www.brebook.com)

Stephen. [from old catalog] Roper, Edwin R. Keller, Clayton W. Pike

Roper's catechism for steam engineers and electricians ..

ROPER'S CATECHISM

FOR

STEAM ENGINEERS

AND

ELECTRICIANS

INCLUDING THE CONSTRUCTION AND MANAGEMENT OF

STEAM ENGINES, STEAM BOILERS AND ELECTRICAL PLANTS

WITH ILLUSTRATIONS

Twenty-first Edition, Rewritten and Greatly Enlarged by

EDWIN R. KELLER, M.E.

AND

CLAYTON W. PIKE, B.S.

PHILADELPHIA:
DAVID McKAY, PUBLISHER,
1022 MARKET STREET

PREFACE TO THE TWENTY-FIRST EDITION.

The great value of a catechism lies in the fact that judicious questioning emphasizes the more important points of a subject and also stimulates the mind of the student to think more definitely and clearly upon the subject than would be the case in merely reading about it. In these respects the written catechism is the best substitute for oral teaching, and the authors trust that this volume will be found of value for this purpose.

The enactment of State laws requiring the licensing of engineers has imposed upon many the necessity of passing examinations for license. The authors likewise hope that it will prove useful to engineers in preparing for such examinations.

EDWIN R. KELLER,
CLAYTON W. PIKE.

PHILADELPHIA, September, 1899.

CONTENTS.

For Alphabetical Index to Subjects, see page 359.

HEAT, FUEL, GASES, WATER, AND STEAM.

HEAT.

THE STEAM ENGINE.

Classification and General Description.

ELECTRICITY.

Fundamental Experiments, Properties, and Units.

Electrical Measurement.

ROPER'S CATECHISM

FOR

STEAM ENGINEERS

AND

ELECTRICIANS.

MECHANICS.

Q. Of what elements are all machines made up?

A. Of six, known as the six mechanical elements. These are the *lever*, *pulley*, *wheel and axle*, *inclined plane*, *wedge*, and the *screw*.

Q. For what is machinery used?

A. To make force available for practical purposes. Machinery does not create force, but transmits it, diffusing it, concentrating it, or changing its direction.

Q. What is force?

A. Force is that which produces motion or tends to produce it. If a force acting on a body meets with a resistance equal and opposite to it, no motion results, but pressure is exerted on the particles of the body. But if the force is not balanced, motion will take place.

Q. What two varieties of force are there?

A. External and internal. External forces are those exerted by bodies on other bodies. Internal forces are those exerted by the particles of a body on neighboring particles. The force of steam against the walls of the pipe or vessel containing it, is external. Each particle of steam exerts an equal amount of force on its neighbor, and this is an example of internal force.

Q. What is the difference between force and pressure?

A. Pressure is a particular case of force. An external force which, on account of a balancing resistance does not produce motion, is generally referred to as a pressure.

Q. What is weight?

A. The weight of a body is the force exerted by the earth on it (an equal amount of force is exerted by it on the earth). When a body rests on another body the upper body exerts upon the lower body a *pressure* or *force* equal to its *weight.* The lower body exerts, of course, an equal and opposite force on the upper.

Q. What is meant by inertia?

A. That property of matter by virtue of which it tends to resist a change of state. Thus, if a body is at rest its inertia makes it offer a resistance to any attempt to put it in motion. If a body is in motion its tendency is to keep moving, and it will do so unless some force is applied to it to bring it to rest.

Q. What is motion?

A. Motion is that property which matter has while it is changing its position.

Q. How would you understand the term *absolute motion?*

A. As a change of position, with reference to some fixed point in space.

Q. What does *relative motion* signify?

A. Change of position, with reference to some other body which we are for the moment considering. Thus two cars in the same train have relative motion with regard to the station which they have left. They have, however, no motion relative to each other.

Q. What is uniform motion?

A. Uniform motion is that in which equal spaces are always passed over in equal amounts of time.

Q. What is variable motion?

A. That in which equal spaces are passed over in unequal amounts of time.

Q. What is accelerated motion?

A. That in which the space passed over in one second is continually increasing or diminishing.

Q. What are Newton's laws of motion?

A. First. A body at rest will remain at rest, or if in motion will continue to move uniformly in a straight line till it is acted upon by some force.

Second. If a body be acted upon by several forces it will obey each, as if the others did not exist, and this will be the case whether the body be at rest or in motion.

Third. If a force act to change the state of a body with respect to rest or motion, the body will offer a resistance equal to and directly opposed to the force. Or to every action there is opposed an equal and opposite reaction.

Q. What is perpetual motion and why is it impossible?

A. See explanation in "Roper's Engineers' Handy-Book," pages 6 and 7.

Q. What is velocity?

A. Velocity is the rate at which motion takes place. If a body moves over a distance of 100 feet in 10 seconds, its velocity is 10 feet per second.

Q. What is uniform velocity?

A. Velocity is uniform when equal spaces are passed over in equal times. If this is not the case the velocity is said to be variable.

Q. What is acceleration?

A. Acceleration is the rate at which the velocity changes, that is, the gain (or loss, as the case may be) in velocity in 1 second.

Q. What case of accelerated motion can you mention?

A. That of a freely falling body which starts from rest, falls 16.1 feet the first second, 48.3 feet the next second, and so on.

Q. What are the simple formulæ which enable us to calculate the performance of falling bodies, when the influence of the friction of the air is considered of no importance?

A. $v = \sqrt{64.4\ h}$ and $h = 16.1\ t^2$.

Q. What is the meaning of the letters in these formulæ?

A. v = velocity in feet per second;
h = height through which the body has fallen, in feet;
t = number of seconds required to fall through the distance h.

Q. If a body falls from a height of 100 feet, what velocity will it have when it reaches the earth's surface?

A. $v = \sqrt{64.4 \times 100} = \sqrt{6440} = 80.2$ feet per second.

Q. How long will it take for the body to fall through 100 feet?

A. $h = 16.1\ t^2$ or $t^2 = \frac{h}{16.1}$; therefore

$$t = \sqrt{\frac{100}{16.1}} = 2.49 \text{ seconds.}$$

Q. What is the acceleration produced by gravity?

A. It is at the surface of the earth, about 32.2 feet per second, and diminishes as we go up from the earth's surface.

Q. What is the mass of a body?

A. It is the quotient of the weight of the body divided by the value of the acceleration due to gravity.

Q. Is the weight of a body everywhere the same?

A. No; it diminishes as we rise from the earth's surface.

Q. Is the mass always the same?

A. Yes; for though the weight changes, the value of the acceleration due to gravity changes to the same extent; therefore the quotient of the two is constant, and this by definition is the mass.

Q. When a force is applied to a body at rest what is the effect?

A. The body is put in motion which is uniformly accelerated. The acceleration produced is proportional to the force, as double the force acting on the same body will produce twice as much acceleration.

Q. If the same force is applied to a body weighing 10 pounds and to another weighing twice as much, on which will it produce the greater acceleration?

A. On the 10-pound body it will produce double the acceleration that it will on the 20-pound body.

Q. What general rule can you give for the relation between force, mass, and acceleration?

A. The force (in pounds) = the mass × acceleration or with sufficient accuracy for most purposes, the force $= \frac{\text{the weight in pounds}}{32.2} \times$ the acceleration in feet per second.

Q. What acceleration will a force of 20 pounds produce if applied to a body weighing 20 pounds?

A. F (force) $= \frac{W\text{ (weight)}}{32.2} \times A$ (acceleration),

$$\text{or } A = \frac{32.2 \times F}{W};$$

$$A = \frac{32.2 \times 20}{20} = 32.2 \text{ feet per second.}$$

This case is that of a freely falling body where the force due to its weight acts upon its mass tending to accelerate it.

Q. What is the momentum of a moving body?

A. It is the force which acting upon it for 1 second will bring it to rest. It is equal to the product of the mass of the body by its velocity.

Q. Has a body at rest any momentum?

A. No; for its velocity is zero, and hence the product of mass times velocity is zero also.

Q. What is work in the science of Mechanics?

A. Work involves two things, *force* and *space*, and the amount of work is equal to the product of force by space. If either is absent no work is done.

Q. What is the unit of work?

A. The foot-pound, which is the amount of work performed in raising a weight of 1 pound through a height of 1 foot.

Q. What example can you give of forces acting without work being done?

A. A weight resting on a table exerts force, but as there is no motion no work is being done by the weight.

Q. Was work done in placing the weight on the table?

A. Yes; if the height of table is 4 feet and the weight is 10 pounds, the amount of work done was 40 foot-pounds.

Q. What is energy?

A. Energy is the power of doing work. For example, the weight on the table has the power to do work if it is allowed to fall from the height of the table.

Q. How many forms of energy are there?

A. Two,—potential energy and kinetic energy. The energy in the weight above mentioned is a case of *potential* energy. A body in motion has also the capacity for doing work stored up in it, and the energy resident in moving bodies is called *kinetic* energy.

Q. Can you give other examples of potential energy?

A. A spring in tension or compression, a tank of water at a height, a reservoir of compressed air, a piece of coal.

Q. Give some examples of kinetic energy.

A. A moving train, a cannon ball, a fly-wheel, a stream of water, the waves of the ocean, heat, electric-current flow.

Q. What is the formula for the energy in a moving body?

A. E (energy in foot-pounds) $= \frac{M \times V^2}{2}$, where M is the mass and V the velocity of the moving body in feet *per second.* In more convenient form, $E = \frac{W \times V^2}{64.4}$, where W is the weight in pounds.

Q. How much energy is stored up in the piston and piston-rod of an engine if the speed of the piston is 600 feet per minute, and their weight is 100 pounds?

A. $E = \frac{100 \times 60 \times 60}{64.4} = 5590$ foot-pounds.

Q. What is the primary source of energy on the earth?

A. The rays of the sun which raise water from sea-level to the clouds from which it falls in rain, and which causes the growth of plants from which has come our coal.

Q. What is the principle of conservation of energy?*

A. That the amount of energy in the universe is fixed and cannot be changed by man. He can transmit it and alter the form in which it appears, as from potential to kinetic, but can in no wise create or destroy it.

Q. What is power?

A. Power is the *rate* at which work is done, or at which energy is changed from one form to another; thus, if a man lifts in one hour 100 weights of 100 pounds each to a height of 4 feet, he has done work at the rate of $100 \times 100 \times 4$, or 40,000 foot-pounds per hour.

Q. What is meant by a horse-power?

A. Doing work at the rate of 33,000 foot-pounds per minute.

Q. In the example above, what horse-power is the man doing?

A. 40,000 foot-pounds per hour $= \frac{40,000}{60}$ foot-pounds per minute, or $666\frac{2}{3}$ foot-pounds per

* See also "Roper's Engineers' Handy-Book," pages 14 and 15.

minute; $666\frac{2}{3} \div 33{,}000 = \frac{2}{100}$ horse-power very nearly.

Q. What is the rule for obtaining the horse-power?

A. To obtain the work done multiply the force in pounds by the distance in feet.

To obtain the power divide this product by the time required to do the work, in minutes.

To obtain the horse-power divide further by 33,000.

Q. How can forces be conveniently represented so as to calculate the effect which they will produce on a body?

A. We represent each force by a line whose direction represents the direction of the force, and whose length is proportional to the amount of the force.

Q. What is the principle known as the *parallelogram of forces?*

A. If two forces acting on a body be represented by two lines forming two adjacent sides of a parallelogram (their lengths being proportional to the strength of the forces and their directions the same as those of the forces), the diagonal of the parallelogram will represent what is called the resultant of the two forces, namely, a force which acting alone would produce on the body the same effect as would the two forces. The direction of the diagonal represents the direction of the resultant or equivalent force, and its length represents the strength of that force.

Q. What is the resultant force which will equal two forces of 3 and 4 pounds, acting at the same point and at an angle of 90 degrees?

A. Lay out the line A B with 4 units of length to represent the force of 4 pounds, and A C with 3 units of length at right angles to A B, to represent the other force. Complete the parallelogram by drawing B D and C D; then the diagonal A D will represent the resultant, and if measured or calculated its length will be found to be 5 units. The resultant force will then be 5 pounds exerted at an angle of 36° 53′ to the line A B.

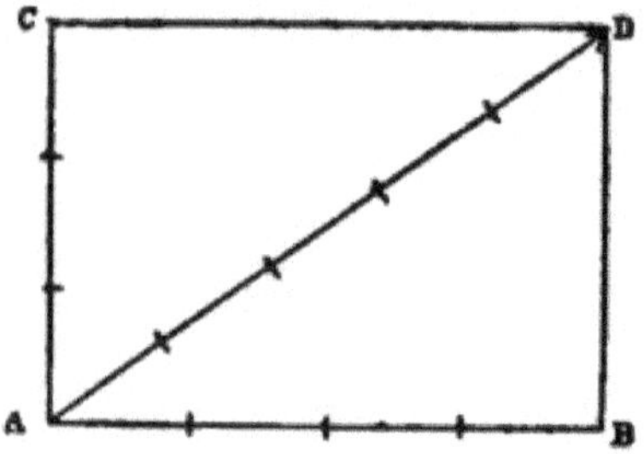

Q. What will be the resultant of a force of 10 pounds in one direction and a force of 5 pounds acting *in the same line* but in the opposite direction?

A. 10 less 5, or 5 pounds. When the forces are parallel or in the same line no parallelogram can be formed.

Q. What is the moment of a force?

A. It is the number which represents its tendency to cause rotation about a certain point. For example, if a stick 5 feet long is pivoted at one end and if a force of 5 pounds be applied at the other end, the force would tend to make the stick rotate about the pivot point. This tendency would be greater if the force were greater or if the length of the stick were greater. It is, in fact, proportional to the product of the force by the perpendicular distance from the pivot point to the line of direction of the force, and this product is technically known as the *moment* of the force about the pivot point.

Q. What is the particular value of the idea of moments?

A. It gives a simple treatment of levers and questions governing the rotation of bodies.

Q. What is the general principle of moments as applied to levers?

A. When two forces are acting at different points in the same body, if the moments, taken about a given point, of the forces are equal and opposed in direction, the body will be at rest, otherwise the body will be set in motion. When there are more than two forces they may be divided into two sets,—one set tending to rotate the body in one direction about the point, and the other set tending to rotate the body in the other direction. If the sum of the moments of the first set of forces is equal to the sum of the moments of the second set, the body will be at rest; but if the sums of the moments of the two sets of forces are unequal, the body will be set in motion.

Q. How does this principle apply to levers?

A. In the use of levers, as, for example, the case of a man trying to raise a rock by means of a crowbar, we have three forces applied at three different points of the crowbar,—one force the strength of the man, another the weight of the rock, and the third the upward thrust of the point of support. By taking moments about the point of support, the moment of the third force becomes zero, since its lever arm is zero, and the bar is in equilibrium under the action of two equal moments. If one force is known, as, for example, the weight of the rock, we can calculate the force which must be applied by the man. If the moment of the force used by the man is the greater, he will move the rock; if less, he cannot do so.

Q. What three classes of levers are there?

A. *First.* Those in which the fulcrum or point of support is between the applied force and the resisting force.

Second. Those in which the resisting force is between the applied force and the fulcrum.

Third. Those in which the applied force is between the fulcrum and the resisting force.

Q. With a lever of the first class, 10 feet long, what force must be applied at the end to lift a weight of 9000 pounds, if the fulcrum is distant from the weight 1 foot?

A. Call the force F. Then by the principle of moments, when the applied force is just sufficient to balance the weight, $F \times 9 = 9000 \times 1$, or $F = 9000 \div 9 = 1000$ pounds.

Q. Is any *power* gained by using a lever, or, more accurately speaking, is any *energy* gained?

A. No; the same expenditure of work is required to raise a weight of 9000 pounds, whatever may be the machinery used to perform the work. A lever merely allows a person, too weak to lift a certain weight with the hands, to do so by taking a longer time to perform the act. Looked at from the standpoint of work, if the 9000 pounds is lifted 1 foot in height, 9000 foot-pounds of work are done. The end of the lever at which the force of 1000 pounds is applied, moves through a distance of 9 feet if the other end moves through 1 foot. Therefore, the work done, which is always the product of force times distance through which the force is exerted, is 1000×9, or 9000 foot-pounds, the same as if the stone were lifted directly.

In one sense it may be said that we gain force by the use of the lever, in that we can, by taking a longer time to do the work, get along with a smaller force.

Q. How does the wheel and axle differ from a lever?

A. The wheel and axle may be considered as a lever in which the points of support and resistance are continually renewed. The center of the axis is the fulcrum, the radius of the wheel is the long arm and the radius of the axle the short arm of the lever.

Q. What is the relation between the applied force and the resulting force in the case of a wedge?

A. If a force of F pounds be applied at the point B in the direction B A, the resulting force W (in a direction perpendicular to A B) will have the following relation:

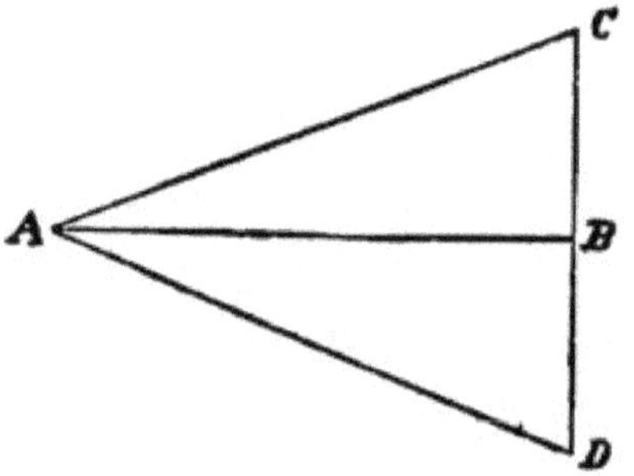

$$\frac{W}{F} = \frac{\text{length A B}}{\text{length C D}}.$$

Q. What two kinds of pulleys are there?

A. The *fixed*, which only turns on its axis, and the *movable*, which moves up and down as well as turns on its axis.

Q. What is the use of a fixed pulley?

A. Merely to change the direction of force.

Q. What advantage is gained by a movable pulley?

A. It enables us to raise a weight by the application of a force half as great as the weight, although we take twice as long to do the work.

Q. With two movable pulleys what would be the gain?

A. We should need a force of only one-quarter the weight.

Q. Does it make any difference whether the movable pulleys are separate or consist of sheaves mounted in the same case?

A. No.

Q. Give the general rule for finding the force necessary to lift a certain weight with the ordinary block and tackle.

A. Divide the weight by the number of sheaves in the movable pulley.

Q. What is the rule for finding the force which must be applied at the end of the lever of a jack-screw in order to lift a certain weight?

A. Multiply the weight by the pitch of the screw, in inches, and divide by 6.2832 times the length of the lever, also expressed in inches.*

*For complete explanation, see "Roper's Engineers' Handy-Book," pages 23 and 24.

POWER TRANSMISSION AND MEASUREMENT.

SHAFTING.

Q. What are the principal methods of transmitting power?

A. By shafting with pulleys and belts.
By rope driving.
By gear wheels.
Hydraulic.
Pneumatic, by compressed air.
Electrical, by dynamos, line, and motors.

Q. Why is shafting now made of steel instead of iron?

A. Because a steel shaft for the same weight and size is stronger with respect to the twisting strain, and stiffer as regards transverse strains due to the weight of pulleys and pull of belts.

Q. What two requirements must be met by shafting?

A. It must be large enough to transmit the required power at the given speed without being twisted too much. It must also have sufficient size to stand the transverse pull due to its own weight, the weight of the pulleys, and the weight and pull of the belts.

Q. What general rule should guide the location of hangers?

A. They should be as near as possible to the pulleys, and should not be over 8 feet apart for light shafting.

Q. Give the rule for calculating the diameter of a shaft to transmit a certain horse-power at a certain number of revolutions per minute.

A. Multiply the horse-power by 70 and divide by the number of revolutions per minute, and extract the cube root of the quotient. The result will be the diameter of the shaft in inches.

Q. What is the rule for obtaining the greatest allowable distance between hangers for a certain size of shaft?

A. Multiply the square of the diameter in inches by 140 and extract the cube root. The result will be the distance in feet.

Q. What is the rule for finding the number of horse-power which a shaft of a certain diameter will transmit at a certain speed?

A. Multiply the cube of the diameter in inches by the number of revolutions per minute and divide the product by 70.

Q. Can these rules be depended upon for all cases?

A. No; only for ordinarily heavy pulleys. For any very heavy pulleys the diameters given by these rules would be too small.

BELTING.

Q. What are the advantages of leather over rubber belts?

A. Leather belts have a longer life, and are less affected by oil and by heat and cold. They will stand being run through shifters or crossed. When worn they can be cut up into narrower belts, whereas rubber belts when worn are of no use.

Q. What two points determine the width of a belt for transmitting a certain horse-power?

A. The speed at which the belt runs and the safe working-strain of the belt, which may be taken as 45 pounds per inch width for single belting.

Q. How much more power will a belt transmit when running at 6000 feet per minute than at a speed of 3000 feet per minute?

A. Twice as much.

Q. At about what speed is it best to run belts?

A. Between 4000 and 5000 feet per minute.

Q. What is a common rule for determining the width of belt to transmit a certain horse-power?

A. That a belt 1 inch wide, at a speed of 1000 feet per minute, will transmit 1 horse-power; a 2-inch belt will transmit 2 horse-power, and so on.

Q. Is this rule a safe one to follow?

A. Yes; for the most favorable cases, where the

belts are open and horizontal, with a long distance between centers, a narrower belt may be used.

Q. Will a belt 30 feet long transmit more power than the same belt 20 feet long?

A. Yes, if it is horizontal; for owing to the greater weight of the longer belt it will sag down a little more in the center and give a little greater arc of contact on the pulleys.

Q. What is the objection to vertical belts?

A. The weight of the belt tends to pull it away from contact with the lower pulley and, therefore, to transmit a given power a vertical belt must be run tighter than if it were horizontal. Moreover, with a horizontal belt the upper side tends to sag down owing to its weight, and this increases the arc of contact with the pulley.

Q. Why do the formulæ of different authors for finding the width of belts differ so much?

A. Because some use a greater permissible tension on the belt than others, which shortens the life of the belt and renders repairs more frequent.*

Q. What is the rule for obtaining the length of an open belt?

A. Multiply the sum of the diameters of the two pulleys by 3.1416 and divide by 2. To the quotient add twice the distance between centers.

*See Belting, "Roper's Engineers' Handy-Book," pages 34–43.

Q. Is this rule strictly accurate?

A. Yes, if the diameters of the pulleys are the same; if not, the result is slightly too small.

Q. How would you measure the length of a belt in a coil?

A. Add the outside diameter to the diameter of the hole and divide by 2. This would give the mean diameter which should be expressed in feet. Then multiply this by 3.1416 and the product by the number of coils in the roll.

Q. How would you determine the proper size of a driven pulley to run at a certain number of revolutions per minute, having given the diameter and speed of the driving pulley?

A. Multiply the diameter of the driver by the number of revolutions which it makes per minute and divide the product by the number of revolutions which the driven pulley is to make.

Q. In arranging for belting, which side should be the loose side, the upper or lower?

A. The upper, so that the weight of the belt may make it sag down and thus make a longer arc of contact between belt and pulleys.

Q. What advantages does rope transmission have over belt driving?

A. The cost of rope is less than that of belting, and the pulleys do not have to be so accurately lined up.

Q. What are the two general methods of using ropes?

A. *First.* To put ropes on like so many parallel spliced belts, one working in each groove of the pulley.

Secondly. To wrap the rope around the pulleys as many times as there are grooves, then to carry it through idlers so arranged that the tension can be varied, and then to carry the rope back to the starting-point and to splice it.

Q. What is the objection to the first method?

A. The separate ropes do not all pull equally.

Q. How is this partially overcome?

A. By making the grooves of the smaller pulley with a sharper angle.

Q. At what speeds do the ropes run?

A. At speeds varying from 25 to 100 feet per second, the most common practice being about 80 feet.

Q. Can you give any figures showing what horse-power is transmitted by a certain size rope?

A. A 1-inch rope at a velocity of 5000 feet per minute will transmit about 13 horse-power.

TOOTHED AND FRICTION GEARING.

Q. What is the *pitch* of a gear wheel?

A. The distance measured along the pitch circle from a point on one tooth to the corresponding point on the next tooth.

Q. What is the *thickness* of a gear tooth?

A. Its width measured along the pitch circle.

Q. What is the *space?*

A. The difference between its *pitch* and its *thickness.*

Q. What is *backlash?*

A. The amount by which the space is greater than the thickness.

Q. What are spur gears used for?

A. To connect parallel shafts.

Q. When are bevel gears used?

A. When it is desired to connect shafts making an angle with each other.

Q. What are the two principal forms of gear teeth?

A. The cycloid and the involute, the latter being used when the number of teeth is small.

Q. How would you calculate the diameter or number of teeth in a driven wheel to run a certain speed having given the diameter or number of teeth of the driver?

A. Just as the diameter of a driven pulley is calculated.*

Q. For what are friction-clutch connections principally used?

A. To take the place of tight and loose pulleys,

*See also "Roper's Engineers' Handy-Book," pages 50–52.

and to connect two or more sections of a line of shafting so that the sections may be disconnected or thrown together without stopping the shaft.

Q. Describe the general principle on which most friction clutches are constructed.

A. A pulley is mounted so as to turn freely on a sleeve in which the shaft turns. This pulley has either a special rim attached to the arms or else a disk attached to the hub, which is gripped between the jaws of the clutch device. The clutch is mounted on and keyed to the shaft. The jaws of the clutch are made to open or shut by moving the clutch collar in one direction or another along the shaft by a fork handle. The motion of the clutch collar operates some kind of toggle joint which moves the jaws; when the jaws are closed so as to grip the rim or disk, the pulley is made to turn with the shaft.

COMPRESSED AIR.

Q. What are some of the purposes for which compressed air is used as a means of transmitting power?

A. For operating cranes, hoists, drills, riveting-machines, coal-mining machinery, railroad signals, shop tools, sand blasts, brakes, etc.

Q. Describe the general method of power transmission by compressed air.

A. Air is compressed by some form of piston pump driven by a steam engine, water wheel, electric motor, or any convenient source of power. Pipes carry the compressed air to the point where it is to be used, where it is led into the air motor or other machine in which it is to be used.

Q. What is the general nature of the air motor?

A. An ordinary steam engine or steam pump may be used as a compressed air motor, according as rotary or reciprocating motion is desired. Commercial motors differ from these only in form and detail.

Q. Why in steam-driven air compressors is the duplex or compound type used so largely?

A. With a single steam and single air cylinder the maximum steam pressure is at the beginning of the stroke, while in the air cylinder the greatest pressure is at the end of the stroke. This is equalized to a great extent by having two cylinders of different sizes and performing the first part of the compression in the larger and finishing it in the smaller cylinder.

Q. Has the compound compressor any other advantage?

A. Yes; it is more efficient, *i. e.*, it compresses a greater quantity of air with a given amount of steam than would a simple compressor.

Q. What is the intercooler?

A. A tank containing coils through which runs cold water. This tank is so connected between the large and small air cylinders that after the air has received the first part of its compression it is led through the intercooler before it passes into the second compressing cylinder.

Q. What is the advantage of the intercooler?

A. The air being cooled after the first compression it does not reach so high a temperature in the second cylinder, so that lubrication is much easier; moreover, it is found that by using the intercooler a given quantity of air can be compressed with the use of a less quantity of steam than would be the case without it.

Q. How much of a saving in steam is attained by the cooling of the air?

A. About 10 per cent. by the intercooler and 5 per cent. by the water jackets around the air-compressing cylinders.

Q. How is the regulation of air pressure maintained?

A. By a balanced valve operating a little piston which in turn operates another controlling the steam supply for the steam cylinder of the compressor.

Q. What are receivers and why are they used?

A. They are steel tanks of suitable size and strength, placed one near the compressor and one near the point where the air is to be used. Their object is to prevent fluctuations of pressure in the system. They thus preserve a steady flow of air in the pipe line and keep the loss of pressure by friction down to a minimum.

Q. What is a common pressure for compressed-air systems?

A. 80 pounds.

Q. How does the loss of pressure due to friction of air flowing through pipes vary?

A. In proportion to the length of pipe and in proportion to the square of the velocity or quantity per minute which goes through the pipe.

Q. Can you give any figures showing the number of cubic feet of compressed air used by air motors?

A. In small motors of, say, one horse-power about 700 cubic feet per horse-power per hour; with large motors as low as 500 cubic feet per horse-power per hour.

Q. What percentage of the power put into the air compressor would you expect to get out of the air motors? In other words, what would be the efficiency of a complete pneumatic transmission system?

A. From 35 to 55 per cent.

ELECTRIC TRANSMISSION OF POWER.

Q. Describe the general method of transmitting power electrically.

A. The energy of a steam engine, water wheel, or other source of power is used to drive an electrical generator or dynamo, which changes energy from the mechanical form into the electrical form. This electrical energy is conveyed from the generator by insulated copper wires of suitable size to the point where it is desired to use the energy. At that point electric motors or other electric devices are attached to the wires and change the energy back again into the mechanical form.

Q. What two classes of transmission are there?

A. Transmission by direct current and transmission by alternating current.

Q. In the electrical transmission of power when would you, generally speaking, use an alternating current transmission, and why?

A. When the distance is over 1500 feet,—because it requires a smaller conductor to transmit a certain power if the pressure used be high than if it be low, and alternating currents can more readily be changed from high to low pressure than can direct currents, and are therefore more convenient to use when high pressures are employed.*

* See also "Roper's Engineers' Handy-Book," page 65.

Q. What three types of direct-current motors are there?

A. The shunt wound, the series wound, and the compound wound.*

Q. For what class of service are these types used?

A. The series motor is used on hoists and street-car motors, where constancy of speed is not necessary, but where a strong starting-torque is desired. The shunt motor is used for the greater part of the work requiring constant speed, the compound motor being used in a few special cases.

Q. What type of direct-current motor is generally used for driving machine tools?

A. The shunt-wound motor, because it naturally runs at nearly constant speed at all loads.

Q. Suppose, as with a lathe, we wish to get several different speeds, how is this accomplished?

A. By a regulating rheostat or controller.

Q. What is the gain, in size of wire used on the line, if we employ a 220-volt system instead of a 110-volt system?

A. The 220-volt system requires but one-quarter the weight of copper in the line.

Q. Do any disadvantages occur to you?

A. The 220-volt line and motor are a little

* For a description of these types see page 300.

more difficult to insulate from the earth, and they are therefore slightly more liable to cause trouble from leakage-currents and accidental shocks.

Q. Is the shock from 220 volts dangerous?

A. Not unless taken by a person in exceedingly delicate health.

Q. Is the shock from 550 volts dangerous?

A. It is exceedingly severe, although rarely, if ever, fatal.

Q. What determines the size of wire to be used for connecting a generator and motor?

A. The power to be transmitted, the pressure used, the distance, and the permissible loss in pressure.

Q. What determines the allowable loss?

A. The variation in speed of the motor, between no load and full load, which you are willing to allow.

Q. Even with no loss of pressure in the line, what variation of speed would you expect with the average small motor?

A. About 3 per cent.

Q. How would you calculate the size of wire, having given the power, pressure, distance, and permissible loss?

A. See "Roper's Engineers' Handy-Book," pages 67, 717, 718.

LUBRICATION.

Q. What is the object of a lubricant?

A. To diminish friction by interposing a thin film between the revolving or sliding surfaces.

Q. Does any lubricant have any tendency to improve a bearing?

A. No; it simply keeps the surfaces apart, diminishes friction and prevents overheating.

Q. What are the requirements for a good lubricant?

A. It must have sufficient body to keep the surfaces apart, but must be as fluid as possible consistent with this requirement. It must have the smallest possible friction, must not gum or corrode; it must have a high flashing-point, and must remain fluid at the lowest temperature at which it will be used.

Q. What would you use for slow speeds and heavy pressures on the bearings?

A. Graphite, soapstone, tallow, or grease.

Q. What is an oil-separator and on what principle does it operate?

A. A device for separating the oil from the steam coming from the exhaust of an engine. The principle on which it operates is to destroy the momentum of the oil which is carried along with the steam. This is accomplished by baffle plates which alter or reverse the direction of flow of the steam. The heavy oil particles are thus thrown against the plates and are given time to fall under the action of gravity into a chamber from which they may be afterward drawn off.

MEASUREMENT OF POWER.

Q. What are three common methods of measuring power?

A. By means of the steam-engine indicator, by electrical methods, and by the Prony brake or some other form of dynamometer.

Q. Which is the most accurate?

A. Whenever the electrical method can be applied it is the quickest and most accurate.

Q. How would you determine by the indicator method the power used by a certain tool?

A. By indicating the engine with the tool running and without it. The difference in the power shown by the two cards gives the power used by the tool.

Q. Is this method accurate?

A. Not if the power used by the tool is small compared to the power of the engine. In this case it is like trying to weigh a fly on a platform scale, by weighing a man on the scale with the fly, and then weighing the man without the fly and subtracting one weight from the other.

Q. What instruments would you require for the electrical method, if direct currents were used?

A. An ampèremeter and voltmeter of proper range or a wattmeter, though the latter is much less commonly at hand.

Q. How would you measure the power used in operating a tool driven by a direct-current electric motor?

A. I would measure the electrical pressure between the two terminals of the motor by connecting to the terminals a voltmeter of suitable range; I would at the same time find what current was supplied to the motor by connecting an ammeter in the circuit suppling the motor; I would take several readings of both instruments and would multiply the average reading of the voltmeter in volts by the average reading of the ammeter in ampères; this product I would divide by 746, and the quotient would be the electrical horse-power supplied to the motor; then I would throw off the belt between the motor and tool and repeat the measurement above so as to get the horse-power used by the motor when running idle; subtracting this from the total power supplied to the motor would give the power used by the tool.

Q. Will this method be correct if the motor is of the alternating current type?

A. No; for the product of volts and ampères does not give the power. In this case a watt-meter must be used.

Q. Describe the Prony brake.

A. The Prony brake consists of two or more blocks of wood attached to a lever arm, and so arranged that they can be clamped more or less tightly to a pulley or shaft, the power transmitted by which it is desired to measure.

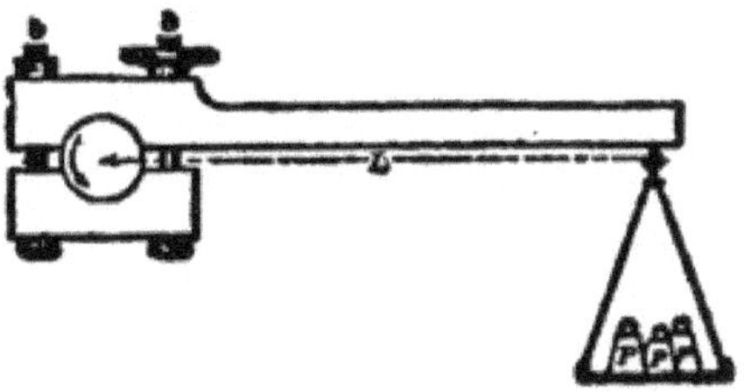

Q. How is the power measured?

A. When the blocks are clamped to the pulley or shaft the tendency is for the Prony brake to revolve with the shaft, but weights are put in the pan hanging from the end of the brake-arm, until this tendency is balanced and the arm stands horizontal. The number of revolutions, *R*, the weight, *W*, and the length, *L*, from the center of the shaft to the point of the lever to which the weight pan hangs, are noted. The horse-power is calculated from the formula—

$$\text{Horse-power} = \frac{W \times L \times R \times 6.28}{33{,}000},$$

or if the distance *L* is made 5′ 3″, the formula becomes, $\text{Horse-power} = \dfrac{W \times R}{1000}$.

Q. What may be substituted for the pan and weights?

A. A spring balance, the average of its readings being used.

Q. What is a dynamometer?

A. Any instrument used to measure power, as, for example, the Prony brake.

Q. For what purpose is a spring dynamometer used?

A. For measuring the power required to propel vehicles, such as carriages, street-cars, or railway coaches.

HEAT, FUEL, AIR, WATER, AND STEAM.

HEAT.

Q. What is heat?

A. Heat is a form of energy. In any body its molecules are in a state of incessant oscillating motion, and the energy of these moving molecules or particles of the body is the heat of that body.*

Q. What is temperature, and how does it differ from heat?

A. Temperature is a measure, not of the heat in a body, but of the tendency of that body to give up its heat to other bodies. Two bodies may be at the same temperature and yet possess very different quantities of heat. For example, a cubic inch of iron and a cubic foot of iron may both be put in the same oven, and after remaining there for a considerable time they would be at the same temperature as would be shown by a thermometer. But the cubic foot of iron has 1728 times as many heat-units in it as the cubic inch, as could be proved by putting them in equal quantities of water, and noting to what temperature the water is raised in each case. According to the molecular theory of the structure of matter a higher temperature means that the molecules of

* For the explanation of the molecular theory of matter, see "Roper's Engineers' Handy-Book," page 611.

the body are moving more rapidly. They, therefore, will communicate motion to surrounding bodies the more readily, and this is the reason that bodies at high temperatures give up heat to those at the lower temperatures. A lower temperature means that the velocity of the molecules is less, and as the temperature gets lower and lower their velocity would become smaller and smaller until a temperature is reached at which their velocity is zero, that is, they are at rest. This temperature is known as the absolute zero of temperatures.

Q. How is temperature measured?

A. By means of a thermometer.

Q. How is a thermometer usually made?

A. A thermometer consists usually of a small hollow glass tube with a bulb at its lower end. The air having been exhausted from the tube it is partially filled with mercury and sealed. The tube is placed in melting ice and the position of the top of the mercury column marked on the glass. The same thing is done with the tube placed in boiling water. The distance between these two marks is divided into a certain number of equal parts, according to which scale is used.

Q. What are the three thermometer scales in common use?

A. The Fahrenheit, Centigrade, and Reaumur.

COMPARISON OF FAHRENHEIT, CENTIGRADE, AND REAUMUR SCALES.

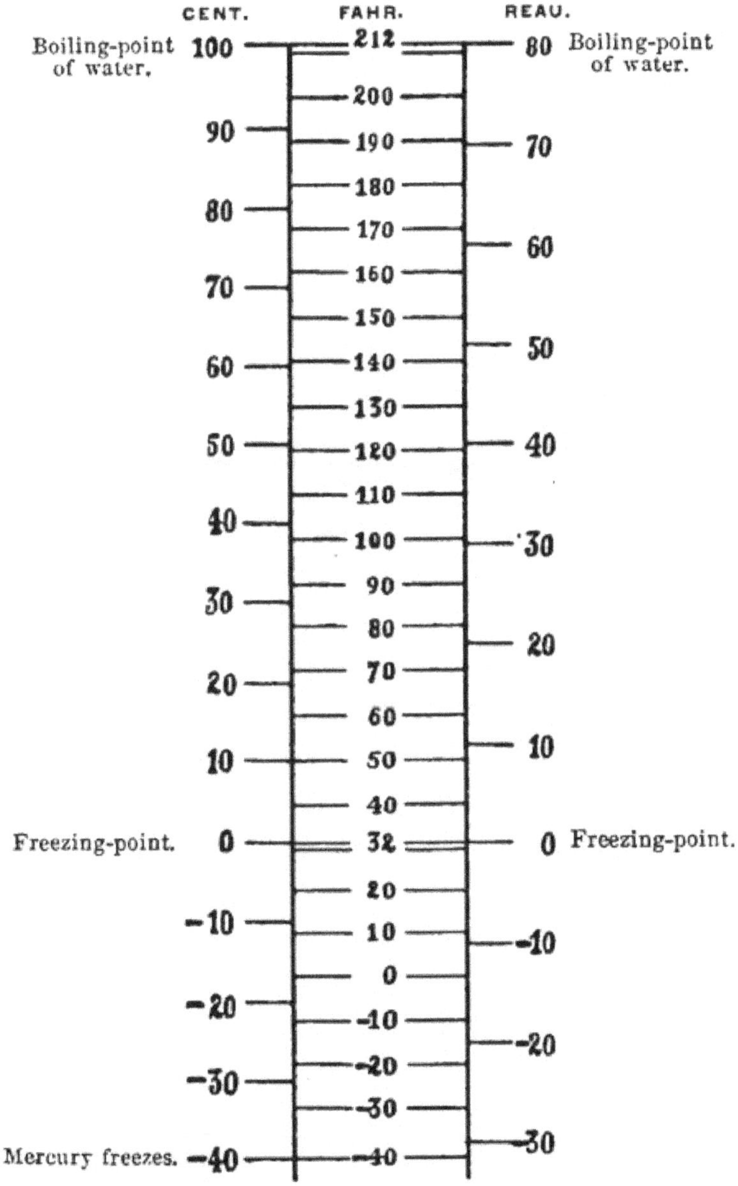

Q. Where is the Fahrenheit scale used?

A. The Fahrenheit scale is used in England, Canada, and in the United States.

Q. What is the difference between Fahrenheit's, Centigrade, and Reaumur's scales?

A. Fahrenheit's zero is 32° below freezing, boiling-point of water, 212°; Centigrade zero is at freezing, boiling-point, 100°; Reaumur's zero is at freezing, boiling-point, 80°. Hence, 180 Fahrenheit degrees are equal to 100 Centigrade degrees or 80 Reaumur degrees, or 9 Fahrenheit degrees are equal to 5 Centigrade or 4 Reaumur degrees.

Q. What are fixed temperatures?

A. One the melting-point of ice, and the other the boiling-point of pure water.

Q. Why do you call these fixed temperatures?

A. Because it is impossible to raise the temperature of ice above 32° Fahr., and no amount of heat will raise boiling water above a temperature of 212° Fahr., if contained in an open vessel.

Q. Does the thermometer indicate the amount of heat in any body?

A. No; only the changes in temperature.

Q. To how high temperatures can the mercurial thermometer be used?

A. To about 600° Fahr. At about 675° mercury vaporizes.

Q. What method is adopted to determine temperatures so high that no thermometer can give a reliable result, as, for example, the temperature in a blast furnace?

A. We take a body, such as platinum, and place a mass of this metal in the blast furnace, and when the mass has acquired the temperature of the furnace we transfer it to a vessel containing a known weight of water. We can then observe the rise of temperature by means of an ordinary thermometer, and from this and the weight of the platinum and its specific heat (.0324) we can calculate the temperature.

Q. What is specific heat?

A. Specific heat of a substance is an expression for the quantity of heat in any given weight of it at certain temperatures. It is the number of heat-units necessary to raise the temperature of 1 pound of the substance 1 degree.

Q. What is sensible heat?

A. That which is sensible to the touch.

Q. What is latent heat?

A. It is that which a body absorbs in changing from a solid to a fluid state, called the latent heat of liquefaction, or that which it absorbs in changing from the liquid to the gaseous state, called the latent heat of vaporization.

Q. What is a unit of heat?

A. The unit of heat is the amount of heat

required to raise the temperature of 1 pound of water 1°, or from 32° to 33° Fahr.

Q. What is the mechanical equivalent of heat?

A. The energy necessary to raise 1 pound 778 feet high; that is, 778 foot-pounds of mechanical energy, if used to produce heat, will be just equal to 1 heat-unit, being just able to raise the temperature of 1 pound of water 1° Fahr.

Q. How is heat transferred from one body to another?

A. In three ways,—by radiation, by conduction, and by convection.*

Q. What substances radiate heat most readily?

A. Those which absorb it most readily and reflect it the least.

Q. What color should the covering of steam pipes be painted?

A. White, because white radiates less than dark colors.

Q. If the pipe is bare, as, for instance, a copper pipe, should it be kept burnished or dull?

A. Burnished.

Q. What are some of the best conductors of heat?

A. Generally speaking, the metals, of which silver, copper, and gold are the best.

* For full explanation, see "Roper's Engineers' Handy-Book," page 94.

Q. Is there any similarity between heat conductivity and electrical conductivity?

A. Generally speaking, good conductors for heat are also good conductors electrically, although the metals do not stand in the same relative order for both cases.

Q. What are some of the best non-conductors?

A. Magnesia, mineral wool, hair felt, cork, air (not in motion).

Q. To what practical use are non-conductors of heat put?

A. To the covering of steam pipes.

Q. Apart from the waste of fuel due to loss of heat by radiation from steam pipes, is there any other effect?

A. Yes; there is a lowering of pressure and a condensation of steam into water, which, if excessive, would cause trouble in an engine.

Q. How much heat does a pound of water receive in passing from a liquid at 212° Fahr. to a vapor at 212°?

A. It receives as much heat as would raise it 966° if the heat was sensible instead of latent.

Q. What is convection of heat?

A. It is the transfer or diffusion of heat in a fluid mass by means of its particles.

Q. Will water boil in a vacuum with less heat than under the pressure of the atmosphere?

A. Yes; in a vacuum of 1 pound absolute pressure water boils at 98° to 100°.

Q. Does water give out heat in freezing?

A. Yes; water in freezing gives 142 heat-units.

Q. What is a thermal unit?

A. It is the quantity of heat required to raise 1 pound of water 1°, the water being at its maximum density (= 39° Fahr.). It is also called a British thermal unit, and is abbreviated B. T. U.

COMBUSTION AND FUELS.

Q. What is combustion?

A. Combustion is a chemical process which takes place rapidly, in which the one or more of the elements which make up the combustible body combines with the oxygen of the air. Briefly, combustion is a rapid oxidation accompanied by flame or fire.

Q. What is smoke?

A. Smoke is the result of imperfect combustion, and its appearance is due to minute unburned particles in the air.

Q. What is necessary to produce complete combustion?

A. We must have sufficient air, must mix the combustible thoroughly with the air, and must maintain the combustible and air mixed with it at a temperature above the igniting-point of the combustible.

Q. What is the meaning of the term fuel?

A. Fuel is used to denote substances that may be burned with air rapidly enough to produce sufficient heat for commercial purposes.

Q. What sort of substances does fuel consist of?

A. Of vegetable substances or the products of their decomposition.

Q. What are some of the principal fuels used in the production of steam?

A. Coal, coke, wood, petroleum, natural gas, peat, and vegetable refuse of various kinds.

Q. What are the elementary substances which are found in most fuels?

A. Carbon, hydrogen, oxygen, nitrogen, and small quantities of other elements.

Q. What is the chief constituent of coal?

A. Carbon.

Q. How much carbon does good coal contain?

A. Anthracite contains about 90 per cent.

Q. Are there any other elements in coal except carbon?

A. Yes; hydrogen, nitrogen, and sulphur in small quantities.

Q. How much heat does 1 pound of pure carbon yield in burning?

A. 14,000 units, approximately.

TABLE

OF TEMPERATURES REQUIRED FOR THE IGNITION OF DIFFERENT COMBUSTIBLE SUBSTANCES.

SUBSTANCES.	Temperature of Ignition.	REMARKS.
Phosphorus, ...	140°	Melts at 110°.
Bisulphide of carbon vapor,	300°	Melts at 130°.
Fulminating powder, . . .	374°	Used in percussion caps.
Fulminate of mercury, . . .	392°	According to Legue and Champion.
Equal parts of chlorate of potash and sulphur, . .	395°	
Sulphur, ...	400°	Melts, 280°; boils, 850°.
Gun-cotton, ...	428°	According to Legue and Champion.
Nitro-glycerine, ...	494°	" " "
Rifle-powder, ...	550°	" " "
Gunpowder, coarse, ...	563°	" " "
Picrate of mercury, lead, or iron, ...	565°	" " "
Picrate powder for torpedoes,	570°	" " "
Picrate powder for muskets,	576°	" " "
Charcoal, the most inflammable willow used for gunpowder, ...	580°	According to Pelouse and Fremy.
Charcoal made by distilling wood at 500°, ...	660°	" " "
Charcoal made at 600°, . . .	700°	" " "
Picrate powder for cannon, .	716°	
Very dry wood, pine, . . .	800°	
Very dry wood, oak, ...	900°	
Charcoal made at 800°, . . .	900°	

It will be seen by the above table that the most combustible substances, generally considered very dangerous, will only ignite by heat alone at a high temperature, so that for their prompt ignition it requires the actual contact of a spark.

Q. How many heat-units does 1 pound of good coal, containing 90 per cent. of carbon, produce?

A. It produces in burning about 13,000 units.

Q. What is the mechanical equivalent of 13,000 units?

A. 10,114,000 foot-pounds,—that is to say, 10,114,000 pounds raised 1 foot high.

Q. How much air does it require to burn 1 pound of coal?

A. About 155 cubic feet.

Q. How much air does it require to burn 100 pounds of coal?

A. About 15,500 cubic feet of air.

Q. What is the difference between anthracite and bituminous coal?

A. Anthracite coal is nearly all carbon, having only about 10 per cent. of other matter, while bituminous coal has from 15 to 50 per cent. of other materials besides pure carbon.

Q. What is the relative fuel value of anthracite coal and wood?

A. A pound of coal is equal to about 2½ pounds of wood.

Q. What is coke?

A. Coke is what is left of coal after the volatile ingredients have been driven off by distillation, as in gasworks; or by partial combustion, as in coke-ovens.

TABLE

SHOWING THE TOTAL HEAT OF COMBUSTION OF VARIOUS FUELS.

SORT OF FUEL.	Equivalent in pure carbon.	Evaporative power in lbs. water from 212° Fahr.	Total heat of combustion in lbs. water heated 1° Fahr.
Charcoal,	0.93	14.00	13,500
Charred peat,	0.80	12.00	11,600
Coke—good,	0.94	14 00	13,620
Coke—mean,	0.88	13.20	12,760
Coke—bad,	0.82	12.30	11,890
COAL:			
Anthracite,	1.05	15.75	15,225
Hard bituminous—hardest,	1.06	15.90	15,370
Hard bituminous—softest,	0.95	15.25	13,775
Coking coal,	1.07	16.00	15,837
Cannel coal,	1.04	15.60	15,080
Long-flaming splint coal,	0.91	13.65	13,195
Lignite,	0.81	12.15	11,745
PEAT:			
Perfectly air-dry,	0.66	10.00	9,660
Containing 25 per cent. water,	..	7.75	7,000
WOOD:			
Perfectly air-dry,	0.50	7.50	7,245
Containing 25 per cent. water,	..	5.80	5,600

REMARK.—In a boiler of fair construction, a pound of coal will convert 9 pounds of water into steam. Each pound of this steam will represent an amount of energy, or capacity for performing work, equivalent to 746,666 foot-pounds, or for the whole 9 pounds, 6,720,000 foot-pounds. In other words, 1 pound of coal has done as much work in evaporating 9 pounds of water into 9 pounds of steam as would lift 300 tons 10 feet high.

Q. Next to carbon, which of the constituents of coal is the greatest heat producer?

A. Hydrogen.

Q. What is the number of heat-units produced by burning a pound of hydrogen?

A. 62,000 British thermal units.

Q. Why do some coals have a greater heat-producing value per pound than does pure carbon?

A. Because they are so rich in hydrogen.

Q. What is meant by the term "free hydrogen" in connection with coal?

A. In all fuel containing carbon, hydrogen, and oxygen, the proportion of hydrogen may be equal to or greater, but never less, than that required to form water with the oxygen. It is only the hydrogen in excess of this which is available as a source of heat, and this is called free hydrogen. The hydrogen existing in combination with oxygen in the state of water, so far from contributing to the actual amount of heat produced, must be evaporated at the expense of the heat developed by the combustion of the carbon.

Q. How does the heat-producing value of petroleum compare with that of coal?

A. It is about ⅓ greater, pound for pound.

Q. What are some of the advantages of using petroleum as a fuel?

A. It gives a steadier fire, is more easily handled, makes no ashes and little smoke, and does not take up so much space.

Q. What determines the advisability of using petroleum rather than coal at a certain place?

A. The most important point is the relative cost of the two.

Q. How many pounds of water can be evaporated by a pound of coal?

A. This depends upon the kind of boiler used and its condition, and also on the kind of coal, the amount varying from 6 to 12 pounds. Under most favorable conditions an evaporation of over 13 pounds of water per pound of combustible has been secured.

Q. What is the meaning of the term "combustible" used in connection with coal; for example, in the expression, "pounds of water evaporated per pound of combustible?"

A. The amount of "combustible" in a quantity of coal is found by subtracting from the original weight of the coal the weight of the water in the coal plus the weight of the ash produced when it is burned.

AIR AND OTHER GASES.

Q. What are the three most important elementary gases—that is, the three most important elements existing naturally in the gaseous state?

A. Oxygen, nitrogen, and hydrogen.

Q. What are some of the most important characteristics of oxygen?

A. It is colorless, tasteless, and odorless. It supports combustion, which process is the chemical combination of the oxygen of the air with the burning substance. It is necessary for the respiration of animals and clearing the blood of impurities. It combines readily with nearly all other chemical elements.

Q. What is iron rust?

A. A combination of iron with oxygen, known as oxide of iron.

Q. What relation does rusting bear to combustion?

A. Rusting is slow oxidation; combustion is rapid oxidation.

Q. What are some of the characteristics of nitrogen?

A. It is also colorless, tasteless, and odorless. Unlike oxygen, it does not combine readily with other elements; it will not burn nor support combustion; mixed with oxygen it forms atmospheric air, its function being to dilute the oxygen.

Q. Give some of the qualities of hydrogen.

A. It is colorless and tasteless and odorless when pure. It is the lightest of known substances, being only one-sixteenth as heavy as air. It unites most readily with oxygen, combining with it to form water in the proportion of 1 part by weight of hydrogen to 8 parts of oxygen. It burns in air with a bluish flame.

Q. Of what does the atmosphere consist?

A. Of oxygen and nitrogen mixed together (not chemically combined), in the ratio of about 1 part by volume of oxygen to 4 parts of nitrogen.

Q. How far from the earth's surface is the atmosphere supposed to extend?

A. At least 45 miles.

Q. Is its density uniform—that is, is it the same at different heights?

A. No; it is less dense as we go farther from the earth's surface.

Q. Does air have any weight?

A. Yes; a cubic foot at the level of the sea weighs about $\frac{8}{100}$ of a pound.

Q. What is atmospheric pressure, so-called?

A. It is the pressure exerted on all bodies by the air, owing to its weight. Since all gases transmit a pressure equally in all directions, and since air has weight, it follows that any square inch of surface has a pressure exerted on it equal to the weight of a column of air 1 square inch in cross-section and of 45 miles or more in length.

Q. How much is this weight, or, in other words, how much is the atmospheric pressure?

TABLE

SHOWING APPROXIMATE INCREASE IN BULK OF AIR DUE TO INCREASE OF TEMPERATURE, AT ATMOSPHERIC PRESSURE.

Fahrenheit.	Bulk	Fahrenheit.	Bulk
Temp. 32 (Freezing-point)	1000	Temp. 75	1099
" 33	1002	" 76 (Summer heat)	1101
" 34	1004	" 77	1104
" 35	1007	" 78	1106
" 36	1009	" 79	1108
" 37	1012	" 80	1110
" 38	1015	" 81	1112
" 39	1018	" 82	1114
" 40	1021	" 83	1116
" 41	1023	" 84	1118
" 42	1025	" 85	1121
" 43	1027	" 86	1123
" 44	1030	" 87	1125
" 45	1032	" 88	1128
" 46	1034	" 89	1130
" 47	1036	" 90	1132
" 48	1038	" 91	1134
" 49	1040	" 92	1136
" 50	1043	" 93	1138
" 51	1045	" 94	1140
" 52	1047	" 95	1142
" 53	1050	" 96 (Blood heat)	1144
" 54	1052	" 97	1146
" 55	1055	" 98	1148
" 56 (Temperate)	1057	" 99	1150
" 57	1059	" 100	1152
" 58	1062	" 110	1173
" 59	1064	" 120	1194
" 60	1066	" 130	1215
" 61	1069	" 140	1235
" 62	1071	" 150	1'55
" 63	1073	" 160	1275
" 64	1075	" 170 (Spirits boil, 176)	1295
" 65	1077	" 180	1315
" 66	1080	" 190	1334
" 67	1082	" 200	1364
" 68	1084	" 210	1372
" 69	1087	" 212 (Water boils)	1375
" 70	1089	" 302	1558
" 71	1091	" 392	1739
" 72	1093	" 482	1919
" 73	1095	" 572	2098
" 74	1097	" 680	2312

A. At sea-level and at 32° Fahr. it is about 14.7 pounds per square inch, or, in round numbers, 15 pounds.

Q. What would you understand by a pressure of three atmospheres?

A. A pressure of 45 pounds per square inch.

Q. What instrument is used to measure atmospheric pressure?

A. The barometer.

Q. How is it made?

A. By filling a glass tube about 3 feet long with mercury and then inverting the tube, letting its open end rest in a vessel containing mercury. The height of the top of the mercury column in the tube is read by a graduated scale.

Q. Why does the mercury not run entirely out of the tube into the vessel?

A. The mercury column is acted upon by two forces; its weight tends to make it run out, but the atmosphere pressing on the surface of the mercury in the vessel resists this action. The mercury column in the tube, therefore, falls only to the point where the pressure per square inch due to the weight of the column is just equal to the pressure per square inch exerted by the atmosphere.

Q. Will the reading of the barometer on a mountain be higher or lower than at sea-level?

A. Lower; for the atmospheric pressure being less, it cannot balance so long a column of mercury.

Q. Why does the mercury column of the barometer at a certain place stand at different heights at different times?

A. Owing to the presence of more or less water, vapor in the atmosphere which changes the weight per cubic foot of air, and consequently alters the atmospheric pressure.

Q. How can the height of a place above sea-level be measured by the barometer?

A. By reading the barometer at the given place and comparing this reading with that taken at some known altitude. Roughly, each inch of length of the barometer column corresponds to a difference in level of 1000 feet.

Q. Can heights also be measured by the thermometer?

A. Yes; by observing at what temperature water boils. At sea-level it boils at 212° Fahr. Roughly, for every 500 feet rise above sea-level the temperature of the boiling-point is 1 degree less.*

Q. What is the effect of heat on air?

A. To expand it.

* For more accurate calculations of heights, see "Roper's Engineers' Handy-Book," pages 121–134.

Q. What is the method of calculating this expansion?

A. Under constant pressure, for each degree Fahr. rise in temperature the volume of air is increased by $\frac{1}{492}$ of its volume at 32° Fahr.

WATER.

Q. Of what is water composed?

A. Of the elementary gases, oxygen and hydrogen, in the proportion by weight of 89 parts of oxygen to 11 parts of hydrogen. By volume the ratio is 1 part of oxygen to 2 parts of hydrogen.

Q. Is pure water found in nature?

A. No; water has, in solution, oxygen, nitrogen, and ammonia, taken up from the air, and traces of salts of many minerals. It may also contain organic impurities resulting from the decomposition of animal or vegetable matter.

Q. Water is taken as the standard for specific gravity of liquids, but is its specific gravity always uniform?

A. No; the weight of a cubic foot of water depends upon its purity. The presence of any salts in solution makes it heavier as in the case of sea water.

Q. Does the temperature of water have any effect upon its specific gravity?

A. Yes; at about 39.2° Fahr. pure water is at

its greatest density, that is, weighs most per cubic foot. Above this temperature it is less dense; below this point it also becomes less dense until at 32° it solidifies into ice.

Q. Under what conditions, then, is water taken as the standard for specific gravities?

A. With the understanding that the water is pure and is at a temperature of 39.2° Fahr.

Q. In what three physical states or forms does water exist?

A. As ice, water, and steam.

Q. How do the weights of a cubic foot of ice, water, and steam compare?

A. A cubic foot of ice weighs about 57 pounds; of water, about $62\frac{1}{2}$ pounds; and of steam, at 5 pounds gauge pressure, $\frac{5}{100}$ pounds, and at 100 pounds gauge pressure, $\frac{26}{100}$ pounds.

Q. What is necessary to change from one of these forms to the other?

A. Merely the application or withdrawal of heat.

Q. Is water a good conductor of heat?

A. No.

Q. Is it a good conductor of electricity?

A. Not if reasonably pure. The addition of some soluble metallic salt, like sodium carbonate or of sulphuric acid, makes it a good electrical conductor.

Q. What are some of its other properties?

A. It is tasteless, odorless, and colorless, and a solvent for most gases and a vast number of liquids and solids.

Q. At what temperature does water boil?

A. This depends upon its purity and upon the atmospheric pressure. Reasonably pure water at the sea-level boils at 212° Fahr.

Q. On a mountain 3000 feet above sea-level, at about what temperature would you expect water to boil?

A. At about 206° Fahr., as for every 500 feet above sea-level the boiling-point drops approximately 1 degree.

Q. How does the boiling-point of salt water compare with that of fresh water?

A. It is higher.

Q. Which will hold the greater quantity of a substance in solution, hot water or cold water?

A. This depends on the nature of the substance. Salts of lime are less soluble in hot water and, therefore, if they exist in a natural water will be deposited when the water is heated to a high temperature.

Q. How does the specific heat of water compare with that of other substances?

A. It is greater than that of nearly all others, and it is for this reason that it is chosen as the standard for specific heats.

Q. What is the specific heat of ice?

A. About .5, or half that of water.

Q. How many units of heat are necessary for melting 1 pound of ice?

A. About 142.

Q. How can water be decomposed into its constituents—oxygen and hydrogen?

A. By passing an electric current through it.*

Q. Can we recombine these two gases to form water?

A. Yes; by burning the hydrogen in a jet in a vessel containing the oxygen.

Q. What is the specific gravity or density of a body?

A. Its weight per unit volume; and since the unit volume used by physicists is the cubic centimeter the specific gravity or density is the weight (in grams) per cubic centimeter.

Q. What would be the specific gravity of pure water?

A. 1, because the weight of a cubic centimeter of pure water is 1 gram.

Q. What is taken as the standard of specific gravities?

A. Water, because its specific gravity is 1.

Q. How could you obtain the specific gravity of any liquid?

*See "Roper's Engineers' Handy-Book," page 134.

A. By weighing equal bulks of the liquid and of water and dividing the weight of the liquid by the weight of the water.

Q. How could you obtain the specific gravity of a solid heavier than water?

A. Weigh it in air; place it in a jar even full of water and catch the overflow of water and weigh it. Divide the weight of the body in air by the weight of the water it displaces; the quotient will be the specific gravity.

Q. When a body whose specific gravity is greater than 1, that is, greater than that of water, is placed in water, what occurs?

A. The body sinks.

Q. How much water does it displace?

A. A volume in cubic feet or inches equal to the volume of the sinking body.

Q. What happens if the specific gravity of the body is less than 1?

A. The body floats, sinking only to a certain depth in the water.

Q. How much water does it displace?

A. Such an amount as will weigh the same as the floating body.

Q. What is meant by the term "head" applied to water?

A. It means a difference in level; for example, with a filled tank at the top of a house, the upper level of the water in the tank being, say, 50 feet above the level of a spigot in the basement, there would be exerted at the spigot a pressure equal in pounds to the weight of a column of water 50 feet high; we should say, then, that there was at the spigot a head of 50 feet.

Q. With a head of 100 feet, how would the pressure compare with the preceding case?

A. It would be double, the pressure being strictly proportional to the head.

Q. What pressure corresponds to a head of 1 foot?

A. Remembering that a cubic foot, or 1728 cubic inches, of water weighs 62.5 pounds, it is easily calculated. A column of water 12 inches high by 1 inch square would contain 12 cubic inches and would weigh $\frac{12}{1728}$ or $\frac{1}{144}$ of 62.5 pounds, or .43 pound. Therefore, the pressure due to a head of 1 foot would be .43 pound per square inch.

Q. When water flows from an orifice in the bottom of a tank under a head, how can its velocity be calculated?

A. Were it not for friction of, and eddy currents in, the water at the orifice, each particle of water would emerge at a velocity the same as it would have if it were allowed to drop through a height equal to the head (the head in this case is the difference in level between the upper surface of the water and the orifice). The formula is $v = \sqrt{64.4\,h}$, or velocity in feet per second equals the square root of 64.4 × the head in feet. Owing to eddy currents set up at the orifice, the actual velocity will be slightly less than the value of v obtained from the formula.

Q. Suppose that you desired to know the number of cubic feet of water flowing from an orifice, how would you obtain it?

A. First obtain, as above, the velocity in feet per second, multiply this by the area of the orifice *in square feet*, and multiply the product by $\frac{6}{10}$. The result will be the quantity in cubic feet per second.

Q. Why do you multiply by $\frac{6}{10}$?

A. Because the jet of water issuing from the orifice has an area less than that of the orifice, it being from six- to eight-tenths as large, according to the form of the orifice.

Q. When water is led from a tank through a long pipe and then allowed to flow from the mouth of the pipe into the air, will the velocity be the same as calculated above?

A. No; it will be less, owing to the friction of the water against the walls of the pipe, which causes a loss of pressure or loss of head.

Q. What does the loss of pressure depend on?

A. The length of pipe, its diameter, and the smoothness of the interior.

Q. Is the loss of pressure greater as the pipe is longer?

A. Yes; the loss is strictly proportional to the length of pipe, the loss for a length of 200 feet being double that for 100 feet.

Q. What effect does increasing the size of pipe have on the loss of pressure?

A. The larger the pipe the less the lost pressure. The loss of pressure is proportional to the length of the pipe and the square of the velocity, and inversely proportional to the diameter of the pipe.*

Q. Having these tables, how would you calculate the velocity at which water escapes from a pipe 500 feet long, the height of the water in the tank being 50 feet above the mouth of the pipe?

A. Calculate first the flow, assuming no loss owing to friction; then, with this flow, from the tables calculate the loss of head; subtracting this head from 50 feet gives the effective head. Finally, using the effective head, calculate the velocity of flow.

* For tables of the loss of pressure, see "Roper's Engineers' Handy-Book," page 42.

STEAM.

Q. What is steam?

A. Steam is the gaseous form of water produced by the application of heat sufficient to raise the temperature of the water to 212° Fahr.

Q. What are the most prominent properties possessed by steam?

A. *First*, its high expansive force; *second*, its property of condensation; *third*, its concealed or latent heat.

Q. Is steam in itself invisible?

A. Yes; and it only becomes visible by loss of temperature, as when a jet is discharged into the open air, and is then seen in the form of vapor.

Q. If a jet of steam flowing into the air gave a cloudy appearance close to the opening, what would you conclude?

A. That the steam was very moist,—that is, that it was carrying along with it a large quantity of water in finely divided particles.

Q. How is the condensation of steam effected?

A. By the lowering of its temperature.

Q. What is the difference in volume between water and steam at a temperature of 212° Fahr.?

A. 1700; that is to say, any given quantity of water converted into steam at the pressure of the atmosphere or 212° Fahr. will present a volume 1700 times greater than its original bulk.

Q. What is dry-saturated steam?

A. The vapor formed from water at a certain temperature and pressure and either remaining in contact with the water, or, if withdrawn from contact with the water, not subjected to any further heating.

Q. What is superheated steam?

A. Dry-saturated steam not in contact with water and raised to a higher temperature than that at which it was formed.

Q. How does ordinary steam differ from dry-saturated steam?

A. It has minute particles of water suspended in it.

Q. Can steam be raised to a very high temperature?

A. Yes; steam can be heated to nearly a red heat, but not while it is held in contact with water.

Q. Is steam at ordinary pressure hot enough to ignite wood?

A. Not without the intervention of some other substance, such as linseed oil, greasy rags, or iron turnings.

Q. What do you understand by the term "steam pressure"?

A. The elastic force which steam exerts in every direction.

Q. What is the sensible heat of steam?

A. The heat which goes to raise its temperature, as, for example, if water at 32° Fahr. has heat applied to it, its temperature will rise up to, but not above, 212° Fahr. The number of heat-units required to raise 1 pound of water from 32° Fahr. to any temperature is called the sensible heat corresponding to that temperature.

Q. What other name is given to the sensible heat?

A. The heat of the liquid or the heat in water.

Q. What is latent heat?

A. Heat which is not sensible to the touch nor indicated by the thermometer.

Q. Is there more than one latent heat?

A. Yes; the *latent heat of liquefaction*, as, for example, the heat absorbed when ice melts into water; and the *latent heat of vaporization*, or the heat absorbed when water is changed to steam.

Q. How may the existence of latent heat be shown?

A. If a thermometer be placed in a vessel containing water which is being heated, the reading of the thermometer increases as heat is applied till it reaches 212°, at which point the water boils. After this, although heat is continually applied, the thermometer goes no higher. This amount of heat which goes to change the physical state of water without changing its temperature is called latent heat.

Q. What is the latent heat of vaporization of water?

A. The amount of heat needed to change a pound of water into steam.

Q. What is the sum of the latent heat of vaporization and the heat of the liquid, at any temperature, called?

A. The total heat corresponding to that temperature.

Q. Is the total heat the same for all pressures?

A. At atmospheric pressure it is 1180, at 100 pounds gauge pressure it is 1217, and at 135 pounds it is 1223.

Q. Does the elasticity of steam increase with an increase of temperature?

A. Yes, but not in the same ratio; because if steam is generated from water at a temperature which gives it the pressure of the atmosphere, an additional temperature of 38° will give it a pressure of 2 atmospheres, and a still further addition of 42° will give it a pressure of 4 atmospheres.

Q. Do you know any simple formula connecting the pressure and temperature of saturated steam?

A. Experiments have been made from which

tables have been constructed, known as tables of the properties of steam, which give the relation between pressure and temperature.*

Q. What is indicated by the ordinary steam gauge?

A. The pressure of the steam above the atmosphere,—that is, the number of pounds by which it exceeds atmospheric pressure.

Q. How would you get the total pressure of the steam,—that is, the number of pounds pressure above zero?

A. By reading the barometer, calculating the number of pounds of atmospheric pressure corresponding to the barometer reading, and adding this to the reading of the steam gauge.

Q. When a pound of steam is condensed to water, how much heat is given up to the surrounding air?

A. An amount of heat equal to the latent heat of steam at the temperature at which it is.

Q. If afterward the water cools to a still lower temperature, how much heat is given off?

A. The amount can be found by subtracting the heat of the liquid at the lower temperature from that corresponding to the upper temperature; the difference will be the number of units of heat given out per pound of cooling water.

*See "Roper's Land and Marine Engines."

THE STEAM BOILER.

Designing steam boilers is not within the province of the stationary engineer. It is his duty not to build boilers, but to operate them to the best advantage. Frequently, however, he is called upon to assist in the selection of the type of boiler for a given purpose, and in this he should remember that the three most important objects to be attained are safety, durability, and economy.

To secure safety it is necessary that the boiler should be made of good material, with good workmanship.

To secure durability the boiler ought to be constructed so as to give the greatest facilities and easiest access for cleaning, repairing, and renewal of any of its parts. The boiler should also be so designed as to avoid unequal strains by expansion and contraction, as far as possible.

In attempting to secure economy in the generation of steam, it is necessary, *first*, to secure perfect combustion of the fuel, so as to produce the greatest amount of heat; *secondly*, to apply the heat in the very best manner to the boiler, so as to heat the water in the most rapid manner possible; *thirdly*, to be very careful to prevent the heat from escaping by radiation or with the products of combustion. If these three conditions be complied with, our arrangements will be of the most economical character. The evaporative efficiency of any boiler and furnace is to be measured by the amount of water evaporated by any given weight of fuel in a given time. Mere waste of fuel, however, is not the only defect attendant upon an inferior construction of boiler and furnace. Where these are not of the best kind, they must be of larger size in order to do the required amount of work; the grate surface must be larger, and more air must be needlessly raised to a higher temperature, thus carrying off a large amount of heat in the waste products of combustion; all of which involves increased outlay of capital and larger running expenses.

Many of the defects of modern boilers might be attributed to the fact that some of the inventors or designers seem to be partly, if not totally, ignorant of the first principles of mechanical science, and to competition between boiler makers themselves, in their efforts to undersell each other; consequently they have to deceive purchasers and steam users by magnifying small boilers into large ones. Therefore, when the boiler comes to be tested, its evaporative powers are found to be lacking, the fuel has to be burned under a sharp draught, and instead of the best results the worst are obtained.

In regard to the metal of the boiler itself, it is a well-known fact that the thicker the iron is, and the poorer its conducting qualities, the greater will be the amount of heat that will be lost or wasted; when, by using a superior quality of iron, one whose tensile strength and conducting powers are both very great, we lessen the resistance to the passage of the heat from the furnace to the water and greatly increase the economy of the boiler. It is well known to engineers that there is a wide difference in the physical properties of different grades of iron and steel used in boiler construction. Some kinds of boiler plate have nearly double the tensile strength of others, and, consequently, to secure the same strength the latter would have to be made twice as thick as the former. This would involve the interposition of a more difficult path between the fire and the water, reducing the efficiency and producing a weaker boiler, because the thicker plate has been subjected to greater strains in the bending. Consequently the thinner plate is by far the more advantageous. On the other hand, as the tensile strength of boiler plates increases, its ductility decreases, and, therefore, great care must be taken in selecting boiler materials, to be sure that they possess not only tensile strength, but also ductility, otherwise the plates will be subjected to initial strains, and, furthermore, the boiler will not be sufficiently flexible to withstand the varying strains to which it is constantly subjected. For these reasons it has been found that the best material for boilers is one which has a moderate tensile strength, 50,000 to 60,000 pounds per square inch, and which will elongate 20 to 25 per cent. before breaking and contract 50 per cent. in cross-section at the point where rupture takes place.

Every attempt to lessen the first cost of a boiler by diminishing the heating- and grate-surface is, to a certain extent, carrying out the principle of "penny wise and pound foolish."

An engine extra large for the work to be done causes a loss of fuel, while a boiler moderately larger than necessary to do the work is productive of economy in the use of fuel. A boiler taxed to its utmost capacity will evaporate, say, from 5 to 6 pounds of water per pound of coal, while the same boiler might evaporate half the quantity of water at the rate of 8 to 10 pounds of water per pound of fuel. This is due partly to the fact that when the boiler is forced the heating surface is not sufficient to utilize all of the heat from the products of combustion, and partly also to the excess of air above that necessary for combustion which passes through the grate and which is heated without producing any useful effect.

For instance, a locomotive boiler burning 10 pounds of coal on each square foot of grate surface in an hour, will evaporate, say, 8 pounds of water for each pound of coal. The same boiler, running at a high speed, and burning 75 pounds of coal on each square foot of grate surface, will evaporate 7 pounds of water for each pound of coal burned. Here is a vast difference in the total amount of evaporation,—each pound of coal produces less steam in the proportion of 9 to 7 pounds.

On the other hand, increasing the size of boiler for a given evaporation must not be carried to excess, because beyond a certain limit there is no advantage to be derived and the increased first cost then becomes a waste in the other direction. There is a certain fixed relation between grate surface, heating surface, and quantity of water evaporated, in each type of boiler, which has been found in practice to be the most advantageous, and any material departure from this in either direction will impair the cost of operation.*

A boiler may generate steam with great economy, but, owing to the steam being wasted by improper application to the engine, the result is unsatisfactory and the boiler unjustly blamed. On the other hand, a boiler that carries out water with its

* For proportions of grate area, heating surface, etc., see page 95 *et seq.;* also, "Roper's Handy-Book," Chapter X.

steam may show a large evaporation, but the steam being wet, is almost useless in the engine; so that in judging the results of a steam-power plant, great care must be taken to examine closely into all of the conditions, before condemning either the boiler or the engine.

In selecting a type of boiler for a given purpose, there are many circumstances to be taken into account. Generally speaking, the most important considerations, as stated above, are safety, economy, and durability; of these, safety should always be first considered, because there are no conditions under which human life and property are not at stake. Consequently, if a boiler is not safe, it is not fit for use under any circumstances. The question of economy must be looked at in a different way. Generally speaking, that boiler is the most economical which evaporates the greatest amount of water with the least consumption of coal, but there may be conditions under which this is not the case; for example, in the coal regions, where fuel is very inexpensive, a highly efficient boiler, which is of necessity more complex than one which is less so, might cost more to operate on account of the interest on the greater first cost and the cost of attendance than a simple flue or even a plain cylinder boiler; and it is a fact that the most efficient and therefore most expensive boilers are not commonly used where fuel is cheap. Similar considerations might lead to the selection of a less durable boiler. Suppose, for example, the case of a bridge to be built in some out-of-the-way locality, the work requiring but a short time and the cost of transportation large compared to the value of the boiler. Under these circumstances it would probably not pay to use a boiler of the highest grade, but preferably one which was merely safe and cheap, did not require much attention, cleaning, etc., and need not necessarily be durable. Such conditions, however, are very uncommon and, generally speaking, the most efficient and durable boiler is the safest and the cheapest.

DIFFERENT TYPES—ADVANTAGES AND DISADVANTAGES.

Q. How would you classify steam boilers?

A. Into cylindrical, flue, fire tubular, and water tubular.

Q. What advantages does the plain cylinder boiler possess over other types?

A. It is simple, inexpensive, easy to clean and repair, and reasonably safe.

Q. What are its disadvantages?

A. Its disadvantages are numerous and great. *First,* on account of its relatively small heating

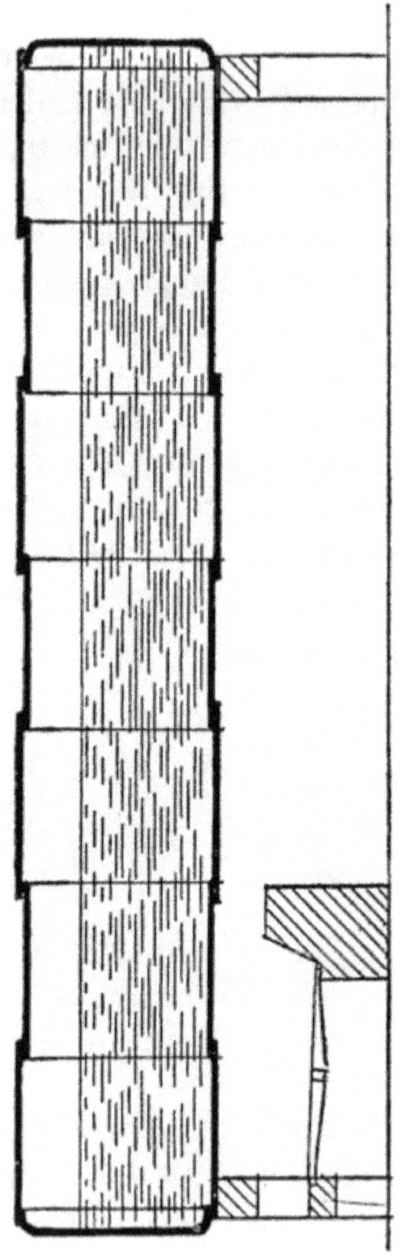

CYLINDER BOILER.

surface, it is very bulky, and, consequently, for a given evaporative capacity, the space it occupies is much greater than in more modern types. *Secondly*, on account of the high temperature at which the gases escape from the stack, it wastes fuel, and for this reason it is the least economical type of boiler in existence. *Thirdly*, it takes a very long time to raise steam. *Fourthly*, the scale formed in the bottom, where the heat is most intense, makes a non-conducting stratum which soon renders that portion of the heating surface useless and causes the iron to burn at that point.

Q. Are plain cylinder boilers much used at the present time?

A. No; they have disappeared almost entirely, mainly on account of their inefficiency. They are found occasionally in localities where the cost of fuel is very low.

Q. Name the principal varieties of flue boilers and briefly describe their characteristics.

A. The Cornish, Lancashire, and Galloway boilers are the principal varieties of flue boilers. In the Cornish type an internal cylindrical flue extends the whole length of the boiler and the furnace is usually contained in the flue. The Lancashire boiler has two internal flues with a furnace in each, the two flues uniting into one

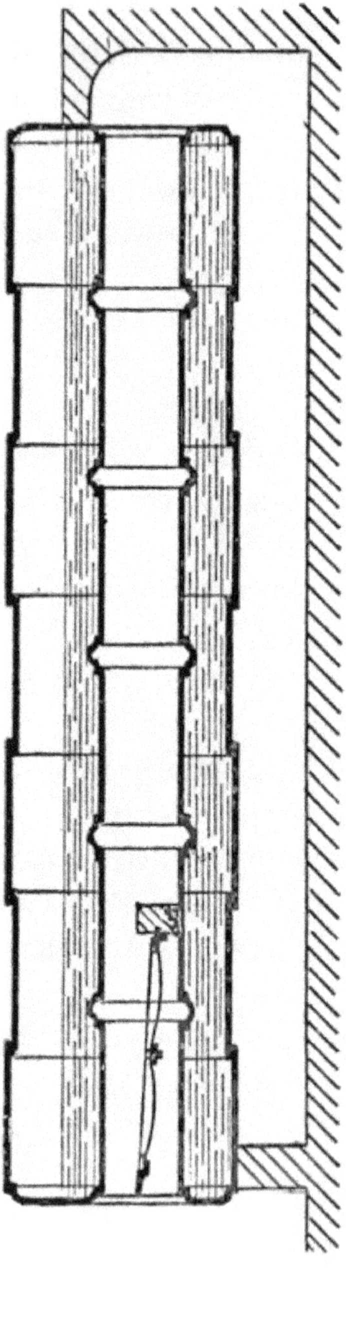

FLUE BOILER, INTERNALLY FIRED.

behind the bridge wall. The Galloway is similar to the Lancashire, but has a number of conical tubes, called Galloway tubes, inside and across the flues, through which the water circulates. The furnaces are either within the flues or external.*

Q. What are the relative advantages and disadvantages of the above-named boilers?

A. The Cornish boiler has a greater heating surface than the plain cylindrical boiler, and it has the further advantage that that portion of the shell on which the scale is deposited, is the coolest instead of the hottest point. It has the disadvantage that, for the same water capacity, it must have a greater diameter.

The Lancashire boiler has the same advantages, and additionally the combustion is more complete than in the Cornish type, because the furnaces may be fired alternately and the smoke which would issue from the stack, if there were but one furnace, is to a great extent consumed by coming in contact with the products of combustion from the other furnace. It also has the disadvantages, in common with the Cornish boiler, that its diameter is greater and, further, the liability of the internal flue to collapse, both of which disadvantages it possesses to an even greater degree than

* For description of flue boilers, see "Roper's Engineers' Handy-Book," pages 160-164.

the Cornish boiler. The liability of the flue to collapse, however, is not very great when the flues are properly stiffened or corrugated.

The Galloway boiler, being virtually a modified Lancashire boiler, possesses all of its advantages; and, additionally, by virtue of the conical tubes, which are placed transversely in the flues, it has a greater heating surface and better circulation. Furthermore, the flues are much less liable to collapse. All of this is accomplished by the Galloway tubes. Of the three boilers mentioned the Galloway type is the safest and most economical in the use of fuel.

Q. What methods are employed to stiffen the flues of boilers and to provide for linear expansion and contraction?

A. This end was formerly accomplished by making the flues in short lengths and connecting them by ∩-shaped rings, riveted on each section of flue. The stiffening of the flue alone is also accomplished by placing T-shaped rings within the flues, at intervals, and by the use of Galloway tubes. This, however, does not take care of expansion and contraction. The best way of accomplishing both ends is by corrugating the flue, which has the further advantage of increasing the heating surface without taking up any more space in the boiler.

Q. What is meant by fire-tube or tubular boilers?

A. Fire-tube or tubular boilers are those in which the combustion gases pass, not only around the outside shell, but also through tubes which are surrounded by water.

Q. In what respect do they differ from flue boilers?

A. In no essential feature, except that instead of one flue of large diameter there are a number of small flues or tubes.

Q. What is the difference between internally and externally fired tubular boilers?

A. The internally fired type consists of an external cylindrical shell containing a furnace extending from the front of the boiler to a point about midway in the length of the boiler. From this point, and extending to the rear end of the boiler, there are a number of tubes which lead the gases of combustion to the back, whence they pass under the outside shell to the front and into the stack. In the externally fired type the tubes extend the whole length of the boiler, and the furnace is outside and under the front end of the boiler. The products of combustion pass along the bottom of the shell to the back of the boiler, and then return through the tubes to the front where they enter the stack connection. From the

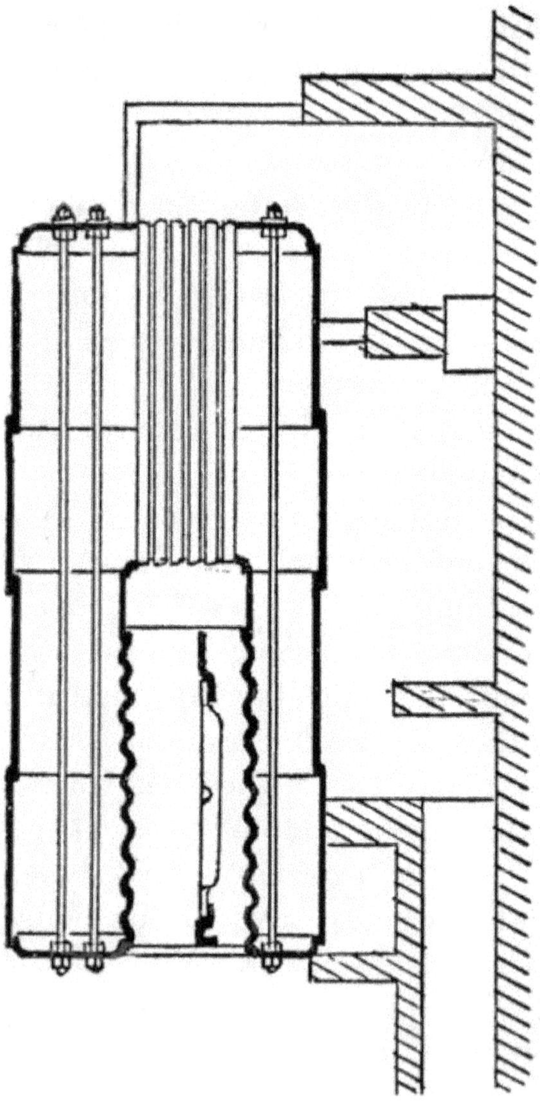

FIRE-TUBE BOILER, INTERNALLY FIRED.

course which the gases take, this latter type is frequently designated as "Return Tubular."*

Q. What advantages does a tubular boiler possess over the cylinder and flue boilers?

A. The tubular takes up less room, generates steam more rapidly, and requires less fuel; moreover, tubes are less dangerous than flues, on account of their small diameter and great strength.

Q. Why are tubular boilers more economical than plain cylinder and flue boilers?

A. Because their heating surface is much greater, and consequently the greater portion of the heat contained in the combustion gases is imparted to the water.

Q. What are their disadvantages as compared to the above-mentioned types? Are they important?

A. The disadvantages are that the first cost is greater, and that they are more difficult to clean and repair, because they are less accessible. These disadvantages are unimportant compared to the great gain in economy.†

Q. What may be said about the tubular boiler in regard to safety?

A. The tubular boiler is just as safe as the cylindrical boiler, and more so than the flue boiler,

* See "Roper's Engineers' Handy-Book," pages 165–168.

† For comparison with water-tube boilers, see next page.

because the parts subjected to internal pressure have the same strength, while those subjected to external pressure, being smaller in diameter, are much stronger.

Q. What is a water-tube boiler?

A. It is one in which the water circulates through a series of tubes, which are surrounded by the combustion gases.

Q. What is the position of the tubes in this class of boilers?

A. Different makers place the tubes in different positions. In the most common type, such as the Babcock and Wilcox, Heine, Gill and Root, the tubes are inclined; in others, such as the Cahall, they are vertical, and occasionally they may even be found curved spirally.*

Q. What are the principal advantages of the water-tube boiler as compared with other types?

A. Its advantages are that it is safer, more economical, steams more rapidly, is easily repaired, more durable; its form may be adapted to almost any existing conditions, and it may be easily taken apart and transported. Its only disadvantages are that it is heavy and expensive.

Q. Why is this type of boiler the most economical in the use of fuel?

* Descriptions of the different types in a condensed form can be found in Babcock and Wilcox's "Steam."

A. Because it has an enormous amount of heating surface, and because the metal which constitutes the heating surface is comparatively light; because the combustion is very thorough, and comparatively little heat is contained in the escaping gases.

Q. Why is it the safest?

A. Because for a given rating the parts subjected to strain are of smaller diameter than in any other type, and, moreover, none are subjected to external pressure. Further, because it is so flexible that the whole structure accommodates itself to changes in temperature without causing undue strains.

Q. What would probably be the difference in an explosion of a water-tube and a fire-tube boiler?

A. Explosions occurring in fire-tube boilers usually wreck the entire boiler, and in some cases whole batteries have been known to explode as the result of a single defect in one of the shells, entailing great loss of life and property. In the water-tube type, while more or less serious explosions have occurred, it is very rare for anything more than a single tube or header to give way; this may be easily repaired and does not generally entail much loss.

Q. Why is it durable?

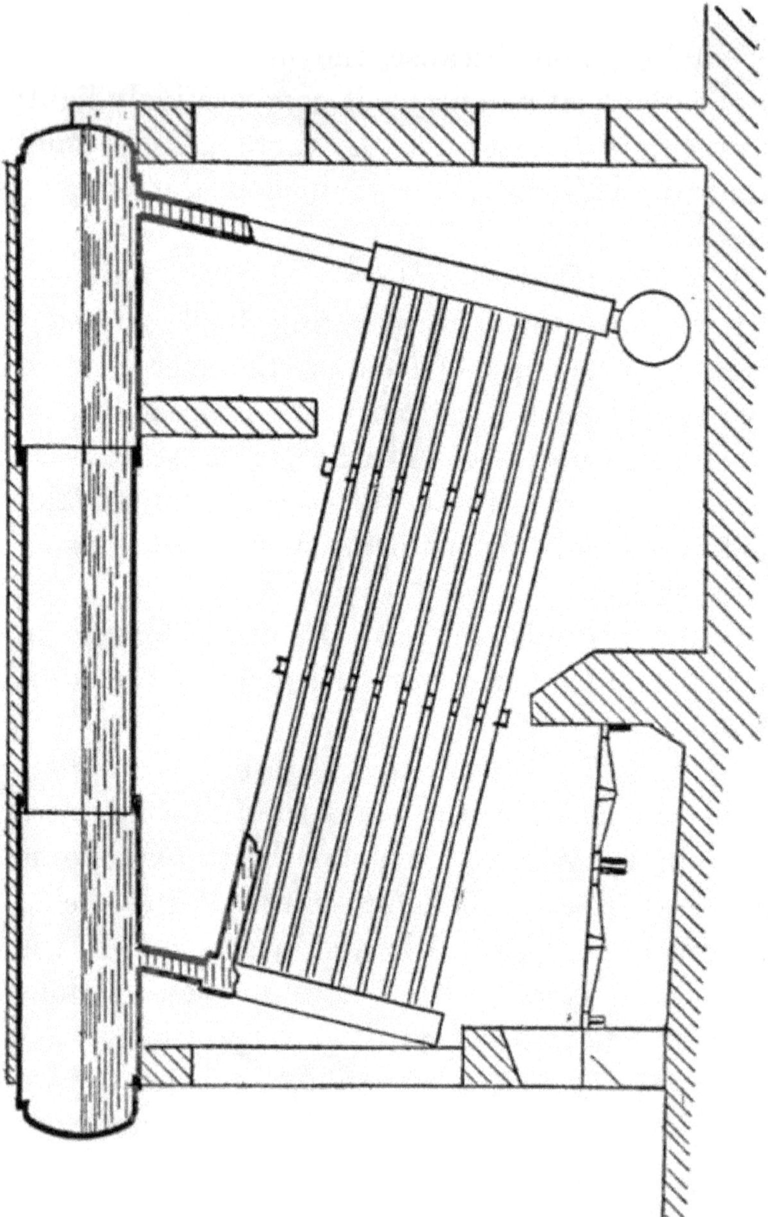

WATER-TUBE BOILER.

A. Because it is easily accessible, and because, as already stated, it adapts itself to the varying expansion and contraction without producing undue strains; further, the circulation is good and consequently the temperature of the different parts is fairly uniform.

Q. To what class do locomotive and marine boilers belong?

A. They may be said to belong to the tubular type, but they have certain characteristics not embodied in the ordinary tubular boiler, which really place them in separate classes by themselves.

Q. Give a brief description of a modern marine boiler.

A. It usually consists of a short, circular shell of large diameter with an internal corrugated furnace. At the back of the furnace is a chamber into which the gases pass from the furnace. This is called the back up-take. A similar chamber in the front, called the front up-take, connects with the stack. The tubes are placed above and around the furnace, and extend from the front to the back up-take.

Q. What, then, is the essential difference between a marine boiler and an internally fired tubular boiler?

A. The principal difference is that while in the

ordinary internally fired tubular boiler the gases pass from the furnaces through tubes to the back and then along the outside to the front; in the marine boiler the gases do not pass around the outside at all, but go from the furnace directly into the back up-take, thence through the tubes to the front up-take and into the stack.

Q. What conditions have brought about this design of boiler for marine purposes?

A. For marine purposes a boiler must be short, as otherwise it could not be set and operated in the available space; and it must be self-contained, because brick setting, on account of its great weight and the motion of the ship, would be out of the question. It must also make steam rapidly.

Q. What pressure may be carried in modern marine boilers?

A. Upward of 200 pounds per square inch.

Q. How many furnaces are generally used?

A. Boilers less than 9 feet in diameter usually have only one; those from 9 to 13 feet, two; over 13 feet, three; and the largest, sometimes exceeding 15 feet in diameter, have four furnaces.

Q. What is meant by a double-ended boiler?

A. When the boilers are fired from the sides of the ship they are frequently placed back to back or are made double-ended—that is, they have furnaces at both ends, with a common or separate back up-takes. The latter arrangement is preferable, because if anything should happen to a tube in one end, this may be repaired without affecting the other half of the boiler.

Q. What are the advantages and disadvantages of marine-type boilers?

A. They do not occupy much floor space, require no brick setting, have a large steaming capacity for a given size and weight, but they are not as economical in the use of fuel or as safe as the best types of land boilers.

Q. Are marine-type boilers ever used for stationary purposes?

A. The marine type of boiler is occasionally found on land. It is well adapted for use where the vibration is so great as to render brick setting impracticable, and where floor space is limited.

Q. Give a brief description of a locomotive boiler.

A. The locomotive boiler consists of a rectangular furnace or fire-box, often made of copper, which contains the grate bars. The fire-box is inclosed in the boiler shell, which is also rectangular where it contains the fire-box, but the remainder of the shell consists of a long cylinder of comparatively small diameter, which contains a large number of tubes. The products of combustion first strike a fire-brick arch which deflects them into the tubes, through which they pass into the funnel or stack placed on the smoke-box at the front end. Locomotives generally use forced draught, which is obtained by allowing the steam from the engine cylinders to exhaust through the funnel.

Q. What conditions have led to the design now generally used for locomotive boilers?

A. A boiler suitable for use on locomotives must be light and of small diameter; light, because it is carried along at a high rate of speed, and of small diameter on account of the limited width of the road bed. For the same reasons, and on account of the jarring motion, brick setting is out of the question, and hence it must be self-contained. It must be capable of making high-pressure steam quickly rather than economically.

Q. Is the locomotive boiler economical in the use of fuel?

A. Yes, but not as economical as the better types of stationary boilers.

Q. How is the necessary strength of the flat surfaces of the fire-boxes obtained in locomotives?

A. By short stay-bolts connected to the outside shell of the boiler. The top of the fire-box is sometimes braced by girders called crown-bars, and sometimes to the semi-circular shell of the boiler above by means of stay-bolts placed radially.

Q. What are the advantages and disadvantages of the locomotive type of boiler?

A. Its advantages are that it is compact, steams quickly, and requires no brick setting. Its disadvantages are that it is expensive, not as economical as the best stationary boilers, and is inaccessible for cleaning and repairs.

Q. Are locomotive-type boilers used for stationary purposes?

A. Yes; they are well adapted for stationary boilers where head room is limited, where it is desired to make steam quickly rather than economically, and where vibration or other conditions would make brick setting undesirable.

Q. How is steam taken from locomotive boilers?

A. Usually from a steam dome placed on the top of the shell. This is to insure dry-steam. Dry-pipes are also sometimes used instead of domes.

HORSE-POWER AND EFFICIENCY.

Q. What is meant by the term *horse-power* as applied to steam boilers?

A. A boiler of one horse-power capacity is one which, under ordinary conditions, supplies as much steam as is consumed in the average steam engine in developing one horse-power.

Q. Is there nothing more definite than this by which the horse-power of boilers may be rated?

A. Yes; the horse-power of steam boilers is now generally based on an evaporative capacity of 30 pounds of water per hour from feed-water at a temperature of 100° Fahr. to steam at a pressure of 70 pounds. This was fixed by a committee of judges at the Centennial Exposition in 1876, and is equivalent to 33,305 heat-units per hour imparted to the water. It is known as the *Centennial Rating.*

Q. How nearly does the horse-power of steam boilers, rated according to this rule, come to the actual consumption of steam in ordinary steam engines?

A. For an automatic cut-off, high-speed, non-condensing steam engine it is just about right. For plain slide-valve engines with throttling governors the Centennial Rating is much too low, while for multiple expansion and condensing engines it is too high.

Q. How, then, would you fix the size of boilers for different engines, assuming that the horse-power of the boilers were based on the Centennial Rating?

A. It is always well to have the boiler capacity a little in excess of that of the engine, because its efficiency is not impaired by operating it below its rated capacity. If the engine were of the high-speed, automatic cut-off, single-expansion, non-condensing type, I should rate the boiler about 10 per cent. higher than the engine; if of the same type, but condensing, about equal; if plain slide valve, non-condensing, with throttling governor, 40 to 50 per cent. higher; the same, condensing, 10 to 20 per cent. higher; if automatic or four-valve non-condensing, about equal; the same, condensing, about 10 to 20 per cent. lower; if compound, high-speed, non-condensing, about 10 per cent. lower; the same, condensing, 15 to 25 per cent. lower; if compound, four-valve, or Corliss, non-condensing, 10 to 15 per cent. lower; the same, condensing, 25 to 35 per cent. lower; if triple expansion, non-condensing, 10 to 15 per cent. lower; the same, condensing, 35 to 45 per cent. lower.

Q. Why are the above rules only approximate?

A. Because the evaporative capacity of a boiler depends on the temperature of the feed-water and also on the pressure of the steam. A boiler of 100 horse-power can evaporate 3000 pounds of water from 100° to steam at 70 pounds pressure; but if the temperature of the feed-water is less, or if the pressure greater, it will not evaporate as much, and *vice versâ.*

Q. What, then, is the best method of determining the size of a steam boiler?

A. The best method is to determine what amount of steam is to be consumed and the pressure at which it is to be delivered to the engine, to specify these requirements and the desired evaporative efficiency to the boiler-maker, and to leave the details of construction to him, binding him to guarantee the boiler to furnish the requisite amount of steam easily and under all conditions.

Q. Approximately, what horse-power of boiler (Centennial Rating) would be required to supply steam to a 100 horse-power, four-valve, non-condensing engine, consuming 26 pounds of steam at 70 pounds pressure per horse-power per hour?

A. Weight of steam required $= 100 \times 26 =$ 2600 pounds per hour; H. P. (Centennial Rating) $= \frac{2600}{30} = 87$, but it would probably be better to use a boiler rated at 90 to 100 horse-power.

Q. What is meant by evaporative efficiency?

A. The number of pounds of steam generated per pound of fuel consumed.

Q. What, roughly, are the results that may be obtained in this respect?

A. In flue boilers of the best types, 6 to 9 pounds; in tubular boilers, 8 to 10 pounds; in water-tube boilers, 10 to 12 pounds of water per pound of coal; the average results, however, are from 10 to 25 per cent. below these figures.

GRATE AREA AND HEATING SURFACE.

Q. What determines the grate surface in boilers?

A. Principally the quality of coal and the draught. In general, it is well to have the grate surface large, but not so large that the air passing through it will be greatly in excess of the amount required for combustion of the fuel.

Q. What amounts of coal can be consumed per square foot of grate surface?

A. Anywhere from 4 to 120 pounds, depending, as already stated, upon the quality of the coal and the draught.

Q. What is meant by heating surface?

A. The heating surface of a boiler means the aggregate area of all of the parts of the boiler which come in contact with the flame or products of combustion on the one side, and with the water or steam on the other. In other words, it is all that part of the surface through which the heat of the fire is transmitted to the water or steam.

Q. How would you calculate the heating surface of different types of boilers?

A. Rule for Cylinder-Boilers.—Multiply $\frac{2}{3}$ of the circumference of the shell in inches by its length in inches, add the area of one end in square inches, and divide by 144. The quotient will be the number of square feet of heating surface.

Rule for Flue-Boilers.—Multiply $\frac{2}{3}$ of the circumference of the shell in inches by its length in inches; multiply the combined circumference of all the flues in inches by their length in inches. Take the sum of these two products and add the area of one end in square inches. Deduct the sum of the areas of the cross-sections of all the flues in square inches. The result divided by 144 is the heating surface in square feet.

Rule for Vertical Tubular Boilers (*such as are generally used for fire-engines*).—Multiply the circumference of the fire-box in inches by its height above the grate in inches. Multiply the combined circumference of all the tubes in inches by their length in inches, and to these two products add the area of the lower tube- or crown-sheet, and from this sum subtract the area of all the tubes, and divide by 144. The quotient will be the number of square feet of heating surface in the boiler.

Rule for Horizontal Tubular Boilers.—Multiply $\frac{2}{3}$ of the circumference of the shell in inches by its length in inches; multiply the combined circumference of all the tubes in inches by their length in inches. To the sum of these two products add $\frac{2}{3}$ the area of both tube-sheets; from this sum subtract the combined area of all the tubes; divide the remainder by 144, and the quotient will be the number of square feet of heating surface.

RULE FOR LOCOMOTIVE BOILERS.—Multiply the length of the furnace-plates in inches by their height above the grate in inches; multiply the width of the ends in inches by their height in inches; multiply the length of the crown-sheet in inches by its width in inches; also the combined circumference of all the tubes in inches by their length in inches; from the sum of these four products substract the combined area of all the tubes and the fire-door; divide the remainder by 144, and the quotient will be the number of square feet of heating surface.

Q. How much heating surface per horse-power should be provided in fire- and water-tube boilers?

A. About 12 to 15 square feet.

Q. How, then, can you approximate the horse-power of a given boiler?

A. By calculating the heating surface in square feet and dividing it by 14.

Q. What is the average ratio between grate and heating surface in stationary boilers?

A. The average is about 35 feet of heating surface to 1 square foot of grate surface. This is for good anthracite coal, but for poorer grades the proportionate surface of the grate should be larger.

Q. How much coal, of good anthracite quality, can be consumed per square foot of grate under ordinary conditions?

A. About 11 pounds.

Q. According to these figures, how much coal, on an average, would be consumed per horse-power per hour?

A. Heating surface per H. P., $= 12$ sq. ft.
Grate " " $= \frac{12}{35}$ "
Coal consumption per sq. ft. of grate per hour, $= 11$ lbs.
Coal consumption per H. P. per hour, $= \frac{12}{35} \times 11 = 3\frac{3}{4}$ lbs.

Q. If all the heat in the fuel were utilized in making steam, what would be the smallest theoretical amount of good anthracite coal consumed per hour?

A. Heat-units required per H. P. (Cent'l R'g), $= 33,305$
Heat-units in best anthracite coal, $= 14,000$
Minimum consumption per H. P. per hour, $= \frac{33305}{14000} = 2.4$ lbs.

BOILER SHELLS.

Q. What materials are used for boiler shells?

A. Wrought iron and steel. The latter is rapidly replacing the former as a boiler material.

Q. Why is steel preferred?

A. Because for a given strength it is lighter; and, as a thinner plate may be used, the efficiency of the heating surface is greater.

Q. What thickness of boiler plate do you consider the safest, most durable, and economical for boilers?

A. *First*, to insure safety in shells and flues of boilers; the thickness proper to use depends very much on the quality of the iron, diameter of boiler, and pressure to be carried. *Secondly*, as to durability, the thickest iron is not always the best, as the outside of the sheet becomes burned and crystallized, and in most cases gives less wear and satisfaction than a thinner gauge. *Thirdly*, as to economy, thin boilers are more economical with fuel, and wear longer, provided in all cases that the diameter and the pressure are in proportion.

Q. What would you consider the proper thickness for boilers?

A. The thickness of boiler iron or steel should range between $\frac{5}{8}$ and $\frac{3}{16}$ of an inch, for the reason that plates of greater thickness than $\frac{5}{8}$ of an inch are liable to burn, especially if the circulation is poor, and they are difficult to work and rivet. If the plates are less than $\frac{3}{16}$ of an inch thick, they cannot be properly caulked, and they are liable to waste away by corrosion so as to impair the safety of the boiler.

Q. What properties should be possessed by materials used for boiler plates?

A. Whether iron or steel, the test-piece should have a tensile strength of not less than 50,000 pounds per square inch; it should elongate 25 per cent. in 8 inches before breaking, and should contract 50 per cent. in cross-section at the point where rupture takes place. It should stand bending without injury around a radius equal to the thickness of the plate.

Q. Is the pressure the same on all riveted seams in boiler shells?

A. No; the pressure on the longitudinal rivets is nearly double that on the curvilinear rivets.

Q. What do you mean by longitudinal and curvilinear rivets?

A. By longitudinal rivets I mean those that run lengthwise on the boiler; the curvilinear are those that are around the circumference of the shell.

Q. If the pressure on the longitudinal seams is double that on the curvilinear, how can all parts of the boiler sustain the same pressure?

A. By making the longitudinal seams double riveted and the curvilinear single.

Q. What is the difference in strength between single- and double-riveted seams?

A. Single-riveted seams are equal to about 56 per cent. of the material used, while double riveting is equal to about 70 per cent.

Q. What do you mean by "equal to about 56 per cent. of material used"?

A. I mean that the boiler plates lose 44 per cent. of their strength in the process of riveting.

Q. What do you consider the proper diameter for rivets of boilers?

A. That would depend very much on the diameter of the boiler, thickness of iron, and pressure to be carried. For boilers from 36 to 42 inches diameter, and $\frac{3}{8}$ iron, if single riveted, the rivets ought to be $\frac{5}{8}$ of an inch for curvilinear, and $\frac{3}{4}$ for the longitudinal; if double riveted, $\frac{5}{8}$ will answer for both longitudinal and curvilinear seams. From $\frac{5}{16}$ iron down to $\frac{3}{16}$ smaller rivets will answer.

Q. Which do you consider the best method of riveting boilers, by hand or by machine?

A. For average or thin boiler plates, hand riveting does very well, but for heavy iron, $\frac{7}{16}$ or $\frac{1}{2}$ inch thick, machine work is far superior; the power of the machine brings the work together better and with less injury to the iron than can be done by hand.

Q. How should the fiber of the iron be placed to give the greatest strength?

A. The direction in which the iron is rolled should always be placed around the boiler, and not lengthwise, because in cylindrical boilers the strain in the line of the axis is much less than the circumferential bursting strain.

Q. Do you consider it an advantage to drill the rivet-holes in boilers instead of punching?

A. Yes; for all first-class work there can be no doubt but that all the rivet-holes ought to be drilled, on account of the liability of the plates to become fractured by the process of punching, causing a great reduction in the strength of the boilers.

Q. Do you consider the use of the drift-pin ought to be dispensed with as much as possible in making boilers?

A. Yes; a reckless use of the drift-pin has in many cases resulted in great injury to the boiler plates; and there is good reason to believe that such injuries as are caused by the drift-pin often hasten the destruction of the boiler.

Q. What is a drift-pin?

A. It is a tapering steel pin introduced into the holes in the seams, to bring them into line.

Q. How do you propose to dispense with the use of the drift-pin?

A. If the holes are laid off carefully in the sheet, and punched with judgment, there will be very little need for the drift-pin, as the holes can be straightened by a flat reamer. Such work will be greatly superior to that where the drift-pin is used.

Q. Do you think it would be of any benefit to slightly heat the boiler plates before rolling them to form the shell of the boiler?

A. Yes; I think it would add very materially to the strength and durability of boilers if the sheets were rolled while warm, as the fiber of the iron would be drawn out; while, in the common practice of cold rolling, the fiber is crushed and broken.

Q. Does hammering improve the quality of iron?

A. No; it only hardens it, but at the same time renders it more brittle, while rolling imparts toughness.

Q. What fact is observable when boiler iron is broken suddenly, as in the case of steam-boiler explosions?

A. It generally presents a crystalline fractured appearance; when, if broken by some slow process, it presents a fibrous or silky appearance,—in the first case the fiber is fractured, and in the other it is drawn out.

Q. What does the crystalline fracture indicate?

A. It indicates hardness, while a fibrous fracture

is a mark of softness and ductility. The finer and more uniform the crystals, the higher the quality of the iron.

Q. Is the pressure equal on all sides of the shell of a boiler when under steam?

A. No; there is more pressure on the lower than on the upper side of a boiler; as the steam presses equally on the surface of the water as on the upper side of the boiler, the weight of the water must be added to the pressure on the lower side.

Q. Are the shells and flues of boilers sometimes injured by the application of the cold-water or "hydrostatic" test?

A. Yes; the shells and flues of boilers are sometimes injured by a reckless use of the test, and in many cases explosions take place soon after the test is applied.

Q. Would the shell and flues of a boiler be stronger under a cold-water pressure of 70 or 80 pounds to the square inch than under the same steam pressure?

A. No; as iron increases in strength by the application of heat up to 550° Fahr., the boiler would be stronger under the steam pressure.

Q. How do you calculate the bursting pressure per square inch of a cylindrical boiler?

A. The rule is to multiply the thickness of the shell in inches by the tensile strength of the material in pounds per square inch, and divide the product by one-half the diameter of the boiler in inches.

Q. How do you calculate the safe working pressure?

A. Multiply the thickness of the shell in inches by the tensile strength in pounds per square inch. Multiply one-half the diameter by the factor of safety. Divide the first product by the second, and the quotient will be the safe working pressure.

Q. What is meant by the factor of safety?

A. By factor of safety is meant the ratio of the ultimate breaking strength to the proper allowable working strength. For example, if a boiler shell is made of steel having a tensile strength of 60,000 pounds and the thickness is calculated with a factor of safety of 4, the greatest strain which would come on any square inch of cross-section is 15,000 pounds; or, in other words, the boiler could carry four times as much pressure before bursting.

Q. What is the factor of safety usually employed in designing boiler shells?

A. It varies from 3 to 5. A safe average for stationary boilers is 4.

Q. What value of tensile strength must be used in the above rules for working and bursting pressure?

A. That depends on how the joints are riveted. The value of tensile strength in the above rules is the ultimate breaking strength of the material multiplied by the efficiency of the joint.

Q. What do you mean by the efficiency of the joint?

A. I mean the number by which the original strength of the material must be multiplied to give its strength after riveting.

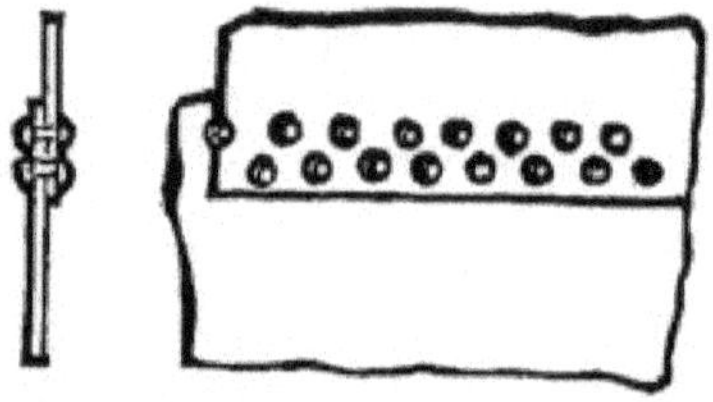

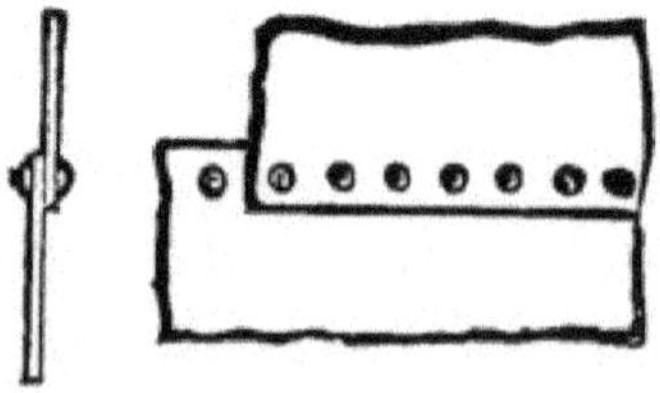

Q. What is the efficiency of single- and double-riveted joints?

A. As already stated above, it is about $\frac{56}{100}$ for single-riveted and about $\frac{70}{100}$ for double-riveted joints. The efficiencies of joints depend, however, not only on the thickness of plate, but also on the spacing of the rivets and the material used.

Q. How would you express by formulæ the relations existing between safe working pressure, bursting pressure, thickness of shell, efficiency of joint, and factor of safety?

A. If p is the safe working pressure in pounds per square inch,
P " bursting pressure in pounds per square inch,
W " ultimate tensile strength in pounds per square inch,
t " thickness of shell in inches,
e " efficiency of the joint,
f " factor of safety,
d " diameter of boiler in inches.

To find the bursting pressure:

$$P = \frac{t \times W \times e}{\frac{1}{2} d}.$$

To find the safe working pressure:

$$p = \frac{t \times W \times e}{\frac{1}{2} d \times f}.$$

To find the thickness of shell for a given working pressure and factor of safety:

$$t = \frac{\frac{1}{2} d \times p \times f}{W \times e}.$$

To find the factor of safety of a given boiler:

$$f = \frac{W \times e \times t}{\frac{1}{2} d \times p}.$$

Q. As an example: If a boiler 48 inches in diameter is made of steel having an ultimate tensile strength of 55,000 pounds per square inch, thickness of shell $\frac{5}{8}$ of an inch, joints double riveted, what is the bursting pressure?

A. $P = \frac{\frac{5}{8} \times 55{,}000 \times .70}{\frac{1}{2} \times 48} = 1000$ pounds per square inch.

Q. With a factor of safety of 5, what would be the safe working pressure?

A. $p = \frac{\frac{5}{8} \times 55{,}000 \times .70}{\frac{1}{2} \times 48 \times 5} = 200$ pounds per square inch.

Q. If the boiler had to work under 150 pounds pressure with a factor of safety of 4, what would be the proper thickness of shell?

A. $t = \frac{\frac{1}{2} \times 48 \times 150 \times 4}{55{,}000 \times .70} = \frac{3}{8}$ of an inch.

Q. If a boiler of the same diameter were made of wrought iron having an ultimate tensile strength of 50,000 pounds, shell $\frac{1}{2}$ inch thick, joints single riveted, what would be the factor of safety for a working pressure of 100 pounds?

A. $f = \frac{50{,}000 \times .56 \times \frac{1}{2}}{\frac{1}{2} \times 48 \times 100} = 5.83,$

which is somewhat higher than is usually allowed by boiler makers.

BOILER SETTING.

Q. What materials should be used for setting boilers?

A. The walls should be of hard burned brick laid in Portland cement. They should be of ample thickness so as to prevent loss by radiation. All surfaces exposed to the action of the hot gases should be lined with best quality fire-brick laid in a thin mortar of fire-clay.

Q. What should be the course of the gases in a tubular boiler?

A. It should be set in such a way that the gases do not pass over the top of the boiler, unless there is ample space for a man to enter and clean off soot.

Q. What should be the distance between the grate bars and the bottom of the boiler shell?

A. Not less than 24 inches. In large boilers it may be as much as 30 inches.

Q. What should be the distance between the back tube sheet and rear wall?

A. From 18 inches for a 48-inch shell to 24 inches for a 72-inch shell.

Q. What is the best method of holding boiler walls in place?

A. With the aid of buck-staves.

Q. What are buck-staves?

A. Vertical cast- or wrought-iron braces placed on the outside of the boiler walls, held together at the top and bottom by tie-rods. Buck-staves are often made of rails, flattened at the end to take the tie-rods.

Q. How should the front of boilers be inclosed?

A. The best method is by a full flush front, which consists of cast-iron plates covering the entire front of the setting, leaving no brickwork in sight. The half-arch front which covers only the furnace is cheaper but less desirable.

Q. When a number of boilers are set together, what plan should be adopted?

A. Each boiler should be set independently of the others, and each should have a separate connection to the stack.

Q. Why is this arrangement better than the old way of setting them in batteries, with a common flue connection?

A. Because each boiler can be operated and shut down independently of the others; because the draught of one is not affected by the others; and, finally, because with the old method of setting, it often happened that when one shell gave out the whole battery exploded.

Q. What kind of boiler should be used where excessive vibration exists or where brickwork would be too heavy?

A. A locomotive- or marine-type boiler is frequently used under these circumstances, because they require no brickwork whatever.

CARE AND MANAGEMENT.

Q. What is the first duty of an engineer when he takes charge of an engine and boiler?

A. It is his duty to examine his boiler and see that the water is at the proper level.

Q. How much water should the boiler contain when in use?

A. The water should be kept up to the second gauge while working, and up to the third at night.

Q. Why should the level of the water be raised at night?

A. As a precaution against the water becoming too low from leakage or evaporation.

Q. In case the water should become dangerously low, what would be the duty of the engineer?

A. He should immediately draw the fire and allow the boiler to cool, and not admit any cold water to the boiler or attempt to raise the safety valve, as it would be positively dangerous.

Q. Why would it be dangerous to raise the safety valve?

A. Because it would lessen the pressure in allowing the steam to escape from the boiler, thus permitting the water to rise up and come in contact with the overheated iron, and probably cause an explosion.

Q. In case the water-supply should be cut off from the boiler for a short time, what would be the duty of the engineer?

A. He should cover his fire with fresh fuel, stop his engine, and keep the regular quantity of water in the boiler until the accident is repaired and the water-supply renewed.

Q. How should an engineer proceed to get up steam?

A. He should first see that the water is at the proper level; he should then remove all ashes and cinders from the furnace, and cover the grate with a thin layer of coal; and after placing wood and shavings on the coal, he will be ready to start the fire.

Q. What advantage is it to place a covering of coal on the grate before the wood or shavings?

A. It is a saving of fuel, as the heat that would be transmitted to the bars is absorbed by the coal, and the bars are also protected from the extreme heat of the fresh fire.

Q. Should an engineer allow his fire to burn gradually when he commences to get up steam from cold water?

A. Yes; as by allowing the fuel to burn very rapidly, some parts of the boiler become expanded to their utmost limits, while other parts are nearly cold. Of course, a great deal depends upon the time in which he has to raise steam.

Q. How should an engineer regulate his fire?

A. He should always keep the fire at a uniform thickness, and not allow any bare places or accumulations of ashes or dead coals in the corners of the furnace, as these places admit great quantities of cold air into the furnace and render the combustion very imperfect.

Q. Should an engineer avoid excessive firing as much as possible?

A. Yes; as excessive firing is always attended with more or less danger, because the intense heat repels the water from the surface of the iron and allows the boiler to be burned.

Q. How thick should an engineer keep his fires?

A. About 3 inches for anthracite coal and about 5 inches for soft coal; but he should regulate the thickness of the fire according to the capacity of the boiler; if the boiler is too small for the engine, the fire should be kept thin, the coal supplied in small quantities and distributed evenly over the grate, and the grate kept as free as possible from ashes and cinders; but if the boiler is extra large for the engine, the thickness of the fire makes but little difference.

Q. What should an engineer do in case, from

neglect or any other cause, his fire should become very low?

A. He should neither poke nor disturb it, as that would have a tendency to put it entirely out, but he should place shavings, sawdust, wood, or greasy waste on the bare places, with a thin covering of coal; then by opening the draught to its full extent the fire will soon come up. If it should become necessary to burn wood on a coal fire, it is always best to make an opening through the coal to the grate-bars, so that the air from the bottom of the furnace can act directly on the wood and increase the combustion.

Q. Should an engineer give great attention to the regulation of the draught in the furnace?

A. Yes; the regulation of draught is one of the most important of an engineer's duties; in fact, it is next in importance to the regulation of the water in the boiler.

Q. How do you explain that?

A. Because it is well known that immense quantities of fuel are recklessly wasted by ignorance and carelessness in the management of the draught.

Q. How should an engineer regulate his draught to obtain the best results from the fuel?

A. He should have no more draught at any time than would produce a sufficient combustion of the fuel to keep the steam at the working pressure, as by opening the damper to its utmost limits great quantities of heat are carried into the chimney and lost.

Q. Can an engineer carry out this principle of regulating the draught in all cases?

A. No; only in furnaces and boilers that are sufficiently large to furnish the necessary amount of steam without forcing. Of course, where the boiler is too small for the engine, or has not sufficient heating surface it is impossible to economize fuel.

Q. Is it objectionable to throw steam or water under the grate-bars of locomotive boilers, when such boilers are used for stationary engines?

A. Yes; as steam or water in the ashpit forms a lye with the ashes and corrodes the iron and destroys the water-legs of the boiler.

Q. Should an engineer in all cases keep his ash-pit clean?

A. Yes; by allowing the ashpit to become filled with ashes and cinders the air becomes heated to a high temperature before entering the fire; the grate-bars also become overheated, and in many cases either badly warped or melted down.

Q. How should an engineer keep his safety valve?

A. He should keep it at all times in good working order, and move it at least once a day, particularly in the morning.

Q. Why should he move the safety-valve every morning?

A. To see that all its parts are in good working order before getting up steam.

Q. Would you consider it reprehensible conduct on the part of an engineer who would weight his safety-valve in order to carry a pressure greater than that he knew to be safe?

A. Yes; such conduct, if proved, ought to be sufficient to disqualify any engineer from ever taking charge of an engine and boiler again.

Q. What is the duty of an engineer in regard to blowing out his boilers?

A. He should carefully remove all the fire from the furnace, and see that the steam is at the proper pressure, say from 45 to 50 pounds. He should also close his damper.

Q. Should any time intervene between the drawing of the fire and the blowing out of the boiler?

A. Yes; at least one hour.

Q. Why should the blowing out of the boiler be deferred for an hour after the fire is drawn?

A. To allow the furnace to cool, and prevent the boiler from being injured with the heat after the water is all blown out.

Q. Why not blow out the boiler under a high pressure of steam, say 70, 80, or even 90 pounds to the square inch?

A. Because the higher the steam pressure the higher the temperature of the iron, so that by blowing out the boiler under a high steam pressure, the change is so sudden that it has a tendency to contract the iron and cause the boiler to leak.

Q. Should the engineer fill his boiler with cold water immediately after blowing out?

A. No; as the introduction of cold water into the boiler before the temperature of the iron becomes lower would in all probability cause the boiler to leak.

Q. How often should an engineer blow out his boiler?

A. Whenever he discovers any appearance of mud in the water.

Q. Is it not customary with some engineers and owners of steam boilers to blow out their boilers once a week?

A. Yes; but the wisdom of this practice is doubtful. When fresh water is boiled, it is supposed to deposit its minerals, and after that it is not advisable to blow out the pure water and fill the boiler with water holding matter in solution and suspension. How often a boiler should be blown out depends on the nature of the water used.

Q. Should an engineer, when filling his boilers, open some cock or valve in the steam room of the boiler and allow the air to escape?

A. Yes; otherwise the air would retard the ingress of the water, and also collect in the steam room of the boiler and prevent the regular expansion of the iron when the fire is started.

Q. What do you mean by the steam room of a boiler?

A. I mean that portion of the boiler occupied by steam above the water.

Q. What is meant by the water room in a steam boiler?

A. That portion of the boiler occupied by water.

Q. What do you call the fire-line of the boiler?

A. The fire-line of the boiler is a longitudinal line above which the fire cannot rise on account of the masonry by which the boiler is surrounded.

Q. How often should an engineer clean the tubes or flues of his boiler?

A. At least once a week; he should also remove all ashes and soot that become attached to the outside of the boiler.

Q. What advantage is gained by cleaning the flues and tubes regularly, and also removing the soot and ashes that become attached to the boiler?

A. It makes a great saving in fuel, as it allows the fire to act directly upon the iron.

Q. How often should an engineer clean his boilers?

A. Every three months, if possible.

Q. Should an engineer, when cleaning his boilers, examine all stays, braces, seams, and angles of the boiler or boilers?

A. Yes; he should make a thorough examination of all parts of the boiler, seams, rivets, crown-sheet, crown-bars, crow-feet, cotters, and braces; he should also sound the shell of the boiler with a very light steel hammer.

Q. Why should the engineer sound the boiler?

A. Because it is the only way in which he can determine the condition of the iron.

Q. How often should an engineer test his steam- or pressure-gauge?

A. At least once a year.

Q. Can an engineer test a steam-gauge himself?

A. No; unless he has a test-gauge, which is not very often the case. The gauge ought to be tested by another gauge built or made expressly for that purpose.

Q. How should an engineer keep his glass water-gauges?

A. He should keep them perfectly clean inside and out.

Q. How can an engineer clean his glass water-gauges inside?

A. By opening the drip-cock and closing the water-valve, and allowing the steam to rush down the glass and carry out the mud or sediment. They should also be swabbed out with a piece of cloth or waste on a small stick, when the boiler is cold; but care should be taken not to touch the inside of the glass with wire or iron, as an abrasion will immediately take place.

Q. In case an engineer has a glass water-gauge, should he neglect his gauge-cocks?

A. No; he should examine them several times in the day, see that they are in good working order, and grind or repair them if necessary. He should always be sure to shut them tight, as by leaving them loose the steam and water destroy the seat of the valve and render them useless.

Q. What evidence do dirty or broken glass gauges, filthy boiler-heads, leaking and muddy gauge-cocks give of a man's ability as an engineer?

A. They furnish strong evidence of his ignorance or neglect of duty.

Q. What should an engineer do in cold weather, when his pumps, boiler connections, steam gauges, or water-pipes are liable to be frozen?

A. He should open all drip- or discharge-cocks and allow the water to run out when he stops work at night, and in the morning make a thorough examination of all steam- and water-connections before he starts his fires.

Q. In case it becomes necessary to stop the engine, and the steam commences to blow off at the safety-valve, what is the duty of the engineer?

A. He should immediately start his pump or injector, and also cover his fire with fresh coal, so that the circulation might be kept up by the feed-water, and the extreme heat of the fire absorbed by the fresh coal, instead of being communicated to the iron of the boiler; and he should not attempt, under any circumstances, to interfere with the free escape of the steam through the safety-valve.

Q. Whenever the fire-door of the furnace is open, should the damper be closed, if possible?

A. Yes; the door and the damper should never be open at the same time, unless it is absolutely necessary, as the cold air, that would otherwise have to pass through the fire and become heated, rushes in through the open door above the fire and impinges on the tube and crown-sheets, and has a tendency to contract the seams and cause leakage.

Q. In case it should become necessary to examine the check-valve while steam is on the boiler, how should it be done?

A. The stop-cock between the check-valve and boiler should be first closed before any attempt is made to unscrew or remove the check. Any neglect to close the stop-cock might result in a serious accident.

Q. How should an engineer proceed to make a joint on the man-hole or hand-holes of his boiler?

A. He should first carefully remove all gum or other material from the seat or flange where the joint is to be made, so that the gasket may have a smooth and solid bearing before he commences to tighten the nut.

Q. Do you know any other important duty an engineer should consider himself bound to perform?

A. Yes; he should daily make a thorough examination of all safety-valves, pumps, injectors, and all steam- and water-connections.

Q. What should be said of an engineer who would allow his boiler and engine to run on from bad to worse, expecting some day to have a general overhauling, instead of making repairs as they were needed?

A. He should be considered totally unfit for the position of an engineer.

Q. When can it be said that an engineer has done his duty?

A. When he shows by his work that he has cared for everything connected with his engine and boiler in the best possible manner.

SCALE-FORMATION, CORROSION, FOAMING, AND PRIMING.

Q. What are the results of scale in boilers, and why?

A. Increased coal consumption and burning of the plates. Because the scale being a poor conductor of heat, the heat of the fire is not imparted to the water as completely as if the scale were not there. For the same reason the water does not protect the iron against crystallization and burning.

Q. What, roughly, is the conductivity of scale as compared to iron?

A. About 1 : 35.

Q. What are the principal ingredients contained in water which cause the formation of scale?

A. Sulphate of lime, phosphate of lime, carbonate of lime, magnesia, silica, and alumina. In sea-water the most important of these is sulphate of lime.

Q. How may the formation of scale be checked?

A. By the use of boiler compounds.

Q. Is there any boiler compound which will be effective in all cases?

A. No; the composition of a boiler compound should be determined by the nature of the impurities. Thus, a proper amount of carbonate of soda introduced regularly with the feed-water would prevent the formation of scale if the ingredient in the water which tends to produce it is sulphate of lime; but this would be of no value if the scale-producing substance is silica or alumina.

Q. What are the principal substances used to check the formation of scale?

A. Carbonate of soda if the scale-forming ingredient is sulphate of lime; phosphate of sodium for the sulphates of lime and magnesium; milk of lime for the carbonates of lime and magnesium; caustic soda and soda ash for the carbonate and sulphate of calcium; and sulphate of magnesium and tannate of soda for the sulphate and carbonate of lime.

Q. How, then, should we proceed if it is found that an undue amount of scale forms in the boiler?

A. We should have a chemical analysis of the feed-water made and add sufficient quantities of the proper kinds of salts to transform the scale-producing ingredients into soluble salts.

Q. In what other ways may the formation of scale be prevented?

A. The use of feed-water heaters and purifiers of the open type is often sufficient, especially where the amount of impurity is not very great.

Q. In what way does this remedy the difficulty?

A. By causing the impurities to be deposited in the heater or purifier, where they can do no harm and whence they may easily be removed without interfering with the operation of the plant.

Q. What is meant by corrosion?

A. By corrosion is meant the wasting, pitting, or grooving of the iron in the boiler.

Q. To what is it generally due?

A. External corrosion is due to the chemical action of sulphur or other products contained in the fuel and in the atmosphere. Internal corrosion is caused by the chemical action of acid and mineral substances contained in the water.

Q. What are the remedies?

A. Numerous remedies are employed to prevent internal corrosion, such as painting the interior of the boiler with Portland cement, allowing a thin layer of scale to form, or suspending metallic zinc in the water and steam spaces, all of which are effective in some cases. There seems to be no effectual remedy against external corrosion when produced by foreign substances contained in the fuel.

Q. What is meant by foaming?

A. By foaming is meant a violent agitation of the water in the boiler. It can be detected by the rising and falling of the level of the water in the gauge glass and by its disturbed condition.

Q. What is the cause of foaming in steam boilers?

A. Foaming in steam boilers might be attributed to different causes. *First*, to the boiler not having a sufficient amount of steam-room, so that whenever the valve opens to admit steam to the cylinder, the pressure on the surface of the water is lessened, allowing the water to rise up from the bottom of the boiler. *Second*, foaming is sometimes caused by the foul condition of the boiler; but in such cases it will be easy to discover the cause, as the water in the glass gauge will appear quite muddy. *Third*, foaming is caused by the presence of any substance of a soapy or greasy nature in the water. But whatever may be the cause of foaming, it is always attended with great danger to the boiler and a certain amount of injury to the engine.

In all cases where a boiler foams badly, the water is lifted from the fire-surface of the boiler, and allows the iron to burn; also, the mud and water from the boiler are carried over with the steam to the cylinder, occupying the clearance between the piston and the head of the cylinder, not only destroying the surface of the cylinder by the grit and dirt, but in many cases causing the fracture of the cylinder-head.

Q. What is the best preventive against foaming?

A. The best preventives against foaming are—*First*, a clean boiler. *Second*, pure water. *Third*, a sufficient amount of steam-room. *Fourth*, a steam pipe well proportioned to the size of the engine.

Q. What is meant by priming?

A. The passage of water from the boiler to the cylinder of the engine in the shape of spray.

Q. How may it be detected?

A. By the appearance of the exhaust from the engine, which, when moist, is white instead of colorless, as is the case when dry, and by a clicking noise in the cylinder, which almost invariably accompanies the presence of moisture.

Q. What causes priming?

A. Usually the want of sufficient steam space in the boiler, or the water being carried at too high a level.

ADJUNCTS OF STEAM BOILERS.

THE SAFETY-VALVE.

The form and construction of this indispensable adjunct to the steam boiler are of the highest importance, not only for the preservation of life and property, which would in the absence of this means of safety be constantly jeopardized, but also to secure the durability of the steam boiler itself.

Increasing the pressure to a dangerous degree would be impossible in any boiler if the safety-valve were what it is supposed to be,—a perfect means for liberating all the steam which a boiler may produce with the fires in full blast, and all other means for the escape of steam closed. Until such a safety-valve shall be devised and adopted in general use, safety from gradually increasing pressure must depend on the attention and watchfulness of the engineer.

Q. What is the object of the safety-valve?

A. It is a valve intended to relieve the boiler from extra pressure, and prevent bursting, collapse, or explosion.

Q. How is this accomplished?

A. By balancing the steam pressure against that of a spring or weight in such a way that when the pressure in the boiler exceeds the limit of safety, it overcomes the action of the spring or weight and opens a valve, allowing the surplus pressure to be relieved.

Q. How often should the safety-valve be moved?

A. At least once a day, more particularly in the morning.

Q. Why should the safety-valve be moved in the morning?

A. So as to be sure that it is in good working order before starting the fire.

Q. What are the most important principles to be adhered to in the construction of the safety-valve?

A. Simplicity of construction, directness, and freedom of action.

Q. Does the safety-valve become worn and leaky by the continual action of the steam?

A. Yes; all safety-valves become leaky and ought to be ground carefully on their seats.

Q. What is the best material to use for grinding safety-valves?

A. Pulverized glass, grit of grinding-stones, or fine emery.

Q. Should safety-valves be constructed with loose or vibratory stems?

A. Yes; as the rigid or solid stem is apt to become jammed by the canting of the lever and weight, and in such cases the higher the pressure the more difficult it is for the valve to open.

Q. What are the principal kinds of safety-valves?

A. There are three principal classes, namely:

(*a*) The dead-weight safety-valve, in which the pressure of the steam is balanced by a weight placed directly on the valve-spindle.

(*b*) The spring safety-valve, which is similar to the above except that the weight is replaced by a spring.

(*c*) The lever safety-valve, in which a weight or spring, instead of acting directly on the valve-spindle, is attached at the end of the lever, the adjustments being made by altering its position on the lever.

Q. What are the relative advantages of springs, as compared to weights in safety-valves?

A. Weights have the advantage that they do not change, which springs are liable to do when in tension. On the other hand, weights could not be used on vessels or locomotives on account of the motion; the momentum which the weight would acquire would constantly alter the blowing-off pressure. For these reasons weight safety-valves are mostly used in connection with stationary boilers, while spring safety-valves are used exclusively for marine and locomotive boilers.

Q. How are safety-valves set for a given blowing-off pressure in the dead-weight and spring type?

A. By simply adjusting the weight or the tension of the spring until it is equal to the blowing-off pressure in pounds per square inch, times the area of the valve in square inches.

Q. How do you calculate what weight should be placed on the end of a given lever safety-valve for a certain blowing-off pressure?

A. Multiply the area of the valve in square inches by the blowing-off pressure in pounds per square inch and the distance of the valve from the fulcrum in inches; multiply the weight of the lever in pounds by the distance of its center of gravity from the fulcrum in inches; multiply the weight of the valve and steam in pounds by their distance from the fulcrum in inches; add the last two products together, subtract their sum from the first product and divide the remainder by the total length of the lever. The quotient will be the required weight in pounds.

Q. How do you calculate the distance of the weight from the fulcrum for a given blowing-off pressure?

A. Multiply the pressure by the area and the distance from the fulcrum from the valve; multiply the weight of the lever by the distance of its center of gravity from the fulcrum; multiply the weight of the valve and stem by their distance from the fulcrum; add the last two products, deduct them from the first product, and divide the remainder by the weight of the ball. The quantities being again taken in pounds and inches, the result will be the distance of the weight from the fulcrum in inches.

Q. How do you calculate the blowing-off pressure for a given position of the ball?

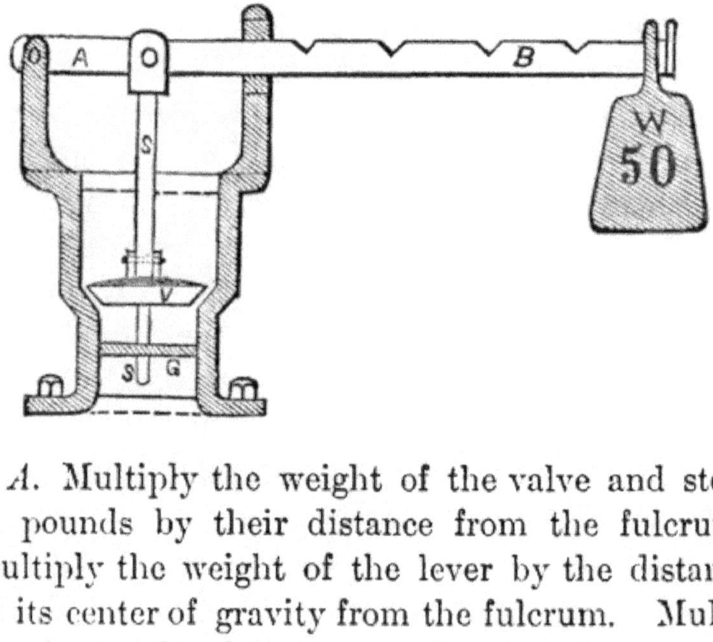

A. Multiply the weight of the valve and stem in pounds by their distance from the fulcrum. Multiply the weight of the lever by the distance of its center of gravity from the fulcrum. Multiply the weight of the ball by its distance from the fulcrum. Multiply the area of the valve by its distance from the fulcrum. Divide the sum of the first three products by the last product. The quantities being all taken in pounds and inches, the result will be the pressure at which the valve will blow off in pounds per square inch.

Q. How can these three rules be expressed by simple formulæ?

A. If in the diagram on opposite page—

W = weight of ball in pounds,
w = weight of valve and stem in pounds,
w_1 = weight of lever in pounds,
l_1 = distance from fulcrum to valve in inches,
l_2 = distance from valve to ball in inches.
l = distance from fulcrum to center of gravity of lever in inches,
p = steam pressure in pounds per square inch,
a = area of valve in square inches,—

then:

$$p\,a\,l_1 = w\,l_1 + w_1\,l + W\,(l_1 + l_2)$$

$$W = \frac{p\,a\,l_1 - [w\,l_1 + w_1\,l]}{l_1 + l_2}$$

$$p = \frac{w\,l_1 + w_1\,l + W(l_1 + l_2)}{a\,l_1}$$

$$l_1 + l_2 = \frac{p\,a\,l_1 - w\,l_1 - w_1\,l}{W}.$$

Q. How would you find the distance of the center of gravity of a lever from the fulcrum?

A. If the lever is of uniform cross-section, as in the diagram shown on page 132, the center of gravity would be at its middle point; but if the lever is taper, proceed according to the following—

RULE *for finding the distance of the center of gravity of taper levers from the fulcrum.*—To the width of the small end of the lever add one-third of the difference, in width, between the large and the small end of the lever. Multiply the sum by the length of the lever, and divide the product by the sum of the large and the small end of the lever, all in inches. The quotient will be the required distance in inches.

Q. How would you express this in a formula?

A. If we let—

a = width of the large end in inches,
b = width of the small end in inches,
l = distance of center of gravity from fulcrum in inches,
L = total length of lever in inches,—

the formula is:

$$l = \frac{a + 2b}{a + b} \cdot \frac{L}{3}.$$

Q. With the aid of this rule and the one given on page 133, find the weight to be placed at the end of the lever of a safety-valve under the following conditions:

width of large end of lever = 3 inches,
width of small end of lever = 2 inches,
total length of lever = 30 inches,
area of valve = 7 sq. inches,
weight of lever = 9 pounds,
weight of valve and stem = 6 inches,
distance of valve stem from
fulcrum = 3 inches,
blowing-off pressure = 60 pounds.

A. By the rule for finding the distance of center of gravity, we have

$$l = \frac{3 + 2 \times 2}{3 + 2} \times \frac{30}{3} = 14 \text{ inches.}$$

By the rule for finding the weight of the ball, we have

$$W = \frac{60 \times 7 \times 3 - [6 \times 3 + 9 \times 14]}{30}$$
$$= 37.2 \text{ pounds}$$

for the required weight to be placed at the end of the lever.

Q. Suppose this weight were moved back so that its distance from the fulcrum became 26 inches, at what pressure would the valve blow off?

A. By the second formula,

$$p = \frac{6 \times 3 + 9 \times 14 + 37.2 \times 26}{7 \times 3} = 53 \text{ pounds.}$$

Q. Where should the weight be placed, so that the valve would blow off at a pressure of 45 pounds?

A. By the third formula,

$$l_1 + l_2 = \frac{45 \times 7 \times 3 - 6 \times 3 - 9 \times 14}{37.2}$$

$$= 21\tfrac{1}{2} \text{ inches from fulcrum.}$$

Q. How large should the area of safety-valves be made for different sizes of boilers?

A. There are a great many rules governing the areas of safety-valves. Some rules base it on the heating surface, some on the grate surface, some on the coal consumption, some on the water evaporated, and some on the heating surface and gauge pressure. The rule given by Professor Thurston gives average values. It is as follows:

RULE. Multiply the heating surface in sq. feet by 5 and divide the product by 10 plus the gauge pressure in pounds per sq. inch. The quotient divided by 2 gives the proper area in square inches.

Q. How much steam should a safety-valve be capable of discharging?

A. About twice as much as that corresponding to the rated capacity of the boiler, because when the boiler is forced to the utmost it is capable of generating a much greater quantity of steam than its rating calls for.

Q. Should a boiler have only one safety-valve?

A. No; it should have at least two, for each boiler fired separately or for each set of boilers placed over one fire.

A TABLE FOR SAFETY-VALVES.

Containing the Circumferences and Areas of Circles from $\frac{1}{16}$ of an Inch to 4 Inches.

Diameter.	Circumference.	Area.	Diameter.	Circumference.	Area.
$\frac{1}{16}$	.1963	.0030	2 ins.	6.2832	3.1416
$\frac{1}{8}$	.3927	.0122	$\frac{1}{16}$	6.4795	3.3411
$\frac{3}{16}$	.5890	.0276	$\frac{1}{8}$	6.6759	3.5465
$\frac{1}{4}$	.7854	.0490	$\frac{3}{16}$	6.8722	3.7582
$\frac{5}{16}$	.9817	.0767	$\frac{1}{4}$	7.0686	3.9760
$\frac{3}{8}$	1.1781	.1104	$\frac{5}{16}$	7.2649	4.2001
$\frac{7}{16}$	1.3744	.1503	$\frac{3}{8}$	7.4613	4.4302
$\frac{1}{2}$	1.5708	.1963	$\frac{7}{16}$	7.6576	4.6664
$\frac{9}{16}$	1.7671	.2485	$\frac{1}{2}$	7.8540	4.9087
$\frac{5}{8}$	1.9635	.3068	$\frac{9}{16}$	8.0503	5.1573
$\frac{11}{16}$	2.1598	.3712	$\frac{5}{8}$	8.2467	5.4119
$\frac{3}{4}$	2.3562	.4417	$\frac{11}{16}$	8.4430	5.6727
$\frac{13}{16}$	2.5525	.5185	$\frac{3}{4}$	8.6394	5.9395
$\frac{7}{8}$	2.7489	.6013	$\frac{13}{16}$	8.8357	6.2126
$\frac{15}{16}$	2.9452	.6903	$\frac{7}{8}$	9.0321	6.4918
			$\frac{15}{16}$	9.2284	6.7772
1 in.	3.1416	.7854	3 ins.	9.4248	7.0686
$\frac{1}{16}$	3.3379	.8861	$\frac{1}{16}$	9.6211	7.3662
$\frac{1}{8}$	3.5343	.9940	$\frac{1}{8}$	9.8175	7.6699
$\frac{3}{16}$	3.7306	1.1075	$\frac{3}{16}$	10.0138	7.9798
$\frac{1}{4}$	3.9270	1.2271	$\frac{1}{4}$	10.2102	8.2957
$\frac{5}{16}$	4.1233	1.3529	$\frac{5}{16}$	10.4065	8.6179
$\frac{3}{8}$	4.3197	1.4848	$\frac{3}{8}$	10.6029	8.9462
$\frac{7}{16}$	4.5160	1.6229	$\frac{7}{16}$	10.7992	9.2806
$\frac{1}{2}$	4.7124	1.7671	$\frac{1}{2}$	10.9956	9.6211
$\frac{9}{16}$	4.9087	1.9175	$\frac{9}{16}$	11.1919	9.9678
$\frac{5}{8}$	5.1051	2.0739	$\frac{5}{8}$	11.3883	10.3206
$\frac{11}{16}$	5.3015	2.2365	$\frac{11}{16}$	11.5846	10.6796
$\frac{3}{4}$	5.4978	2.4052	$\frac{3}{4}$	11.7810	11.0446
$\frac{13}{16}$	5.6941	2.5801	$\frac{13}{16}$	11.9773	11.4159
$\frac{7}{8}$	5.8905	2.7611	$\frac{7}{8}$	12.1737	11.7932
$\frac{15}{16}$	6.0868	2.9483	$\frac{15}{16}$	12.3700	12.1768
			4 ins.	12.5664	12.5664

GAUGES.

Q. What is meant by a gauge?

A. A gauge is any instrument or device used for measuring.

Q. What are the principal gauges used in connection with steam boilers?

A. The steam pressure gauge, vacuum gauge, water gauge, salinometer, and econometer.

Q. Describe the steam gauge.

A. There are two kinds: those which merely indicate the pressure and those which make a permanent record of it. Both are usually constructed on the principle invented by Bourdon, and consist of a thin tube of elliptical cross-section, bent into a curved shape. The steam whose pressure is to be measured is admitted into the tube and tends to make the cross-section circular. This tendency causes the tube to straighten itself out partially, and the instrument is so constructed with a pointer and gearing that the straightening of the tube moves the pointer which indicates the pressure within on a suitable dial. The recording gauge has, in addition, a clock which moves the dial, giving it one revolution in 24 hours, so that by the aid of a pen or stylus filled with ink a complete record of the pressure carried during this time can be had.

Q. Do steam gauges register absolute pressure?

A. No; they are usually constructed to indicate pressure above the atmosphere—that is, at atmospheric pressure (14.7 pounds per square inch) the pointer stands at zero.

Q. What precautions should be taken in using pressure gauges?

A. The pointer should always stand at zero when there is no pressure in the boiler. If it does not, it should be adjusted. Even after this is done, the readings at other pressures may be incorrect and its readings should be checked from time to time by comparing with a standard gauge which is known to be correct.

Q. What is a vacuum gauge?

A. It is made in the same way as a pressure gauge, but it is arranged to read pressures below the atmosphere instead of above.

Q. How are vacuum gauges calibrated?

A. They are usually calibrated in inches of mercury instead of pounds,—that is to say, the readings indicate to how many inches the vacuum would allow a column of mercury to rise under atmospheric pressure. Each inch of mercury corresponds roughly to a vacuum of about half a pound, so that a reading of 20″ on a vacuum gauge would mean that the pressure is about 10 pounds below that of the atmosphere.

Q. Why are they calibrated in this way and not in absolute pressures?

A. Because the mechanism which operates the gauge depends for its action upon the *difference* in pressure of the atmosphere and vacuum chamber; hence, as the pressure of the atmosphere varies, the gauge would not be accurate if calibrated in pounds absolute pressure.

Q. What is a water gauge?

A. It is a device for indicating the level of the water in the boiler. It usually consists of a plain glass tube placed on the outside of the boiler, and connected at the top to the steam- and at the bottom to the water-space.

Q. What is a safety water column?

A. It is a modification of a glass water gauge, with floats so arranged that a signal is given both when the water is too high and when it is too low.

Q. Do you consider the use of safety water columns advisable?

A. They are very useful where an engineer or fireman has other duties to perform besides attending to the boiler; but it is a mistake for engineers to neglect watching the water-level on account of this device because it may get out of order. There can be nothing so dangerous in running boilers as neglecting the water. In some instances where these safety water columns were used, the firemen have been known to systematically fall asleep and depend on the alarm in the safety water column to awaken them at the proper time.

Q. Is the glass gauge the only device used for indicating the water-level?

A. No; every boiler should, in addition, be fitted with gauge cocks placed at different levels. These are partly for the purpose of checking up the glass gauge and partly for use in case the gauge glass should break, which is not an infrequent occurrence.

Q. What is the salinometer?

A. It is an instrument or gauge used for indicating the quantity of salt contained in the water of marine boilers.

Q. What is the econometer?

A. It is an instrument or gauge used for indicating, continuously and automatically, the quantity of carbonic acid contained in the products of combustion.

Q. How much carbonic acid should they contain?

A. As much as possible.

Q. How can this be attained?

A. By supplying enough air to the furnace for a complete combustion of the fuel, but not much in excess of that amount.

Q. What is the result if too much air is supplied?

A. A portion of the heat of combustion is consumed in raising the temperature of the excess of air and consequently wasted. The following table shows the amount of wasted fuel for different percentages of carbonic acid in the flue gases:

TABLE

SHOWING WASTE OF FUEL DUE TO EXCESSIVE SUPPLY OF AIR.

(COAL OF MEDIUM QUALITY.)

Percentage carbonic acid in flue gases, ...	2	4	6	8	10	12	14
No. of times the quantity of air required for complete combustion,.	9.5	4.7	3.2	2.4	1.9	1.6	1.4
Percentage waste of fuel at 420° Fahr.,.	90	45	30	23	18	15	13

PUMPS AND INJECTORS.

Q. What is a pump?

A. It is a device for lifting, forcing, or transferring water or other liquids.

Q. How are pumps usually operated?

A. (*a*) By belting or gearing from some power shaft, called power pumps.

(*b*) By the direct connection to a steam cylinder equipped with suitable valve gear for the distribution of the steam, called steam pumps.

(*c*) By direct connection or gearing to an electric motor; these are called electric pumps.

Q. Which of the above types is usually adopted for feeding boilers?

A. The steam pump.

Q. What different kinds of steam pumps are there?

A. (*a*) Fly-wheel pumps—those in which the reciprocating motion of the steam piston is first converted into rotary motion by means of a crank shaft, with a fly-wheel to help it over the dead centers, and then re-converted by another crank and rods into reciprocating motion for the water cylinder.

(*b*) Direct-acting pumps—those in which the water piston or plunger is mounted on the same rod as the steam piston and the power transmitted from the latter to the former, direct and without the intervention of a crank shaft and fly-wheel. In this type an auxiliary valve gear is required in addition to the main valve gear, to help the machine over its dead points.

(*c*) Duplex pumps—consisting of a combination of two pumps so coupled together that the steam-valve of the one is operated by the piston of the other, and *vice versâ.*

Q. Which of these is most commonly used as a boiler-feed pump? Why?

A. The duplex pump, because it is the simplest.

Q. What is the difference between a force pump and a suction pump?

A. A force pump is one in which the energy is expended in forcing the water against some opposing pressure, such as that in the boiler. A suction pump is one which takes the water from a lower level than that of the pump, as, for example, a pump placed at the top of a well.

Q. Is there any limit beyond which water cannot be lifted by a suction pump? Give reasons.

A. Yes; water cannot be lifted by a suction pump over 33 feet vertically, and it will deliver water slowly only, at this height. The reason for this is that the pump does not actually lift the water, but merely creates a vacuum in the water cylinder, and the water is lifted by the atmospheric pressure on its surface. The atmospheric pressure will support a column of water about 33 feet in height, hence this is the limit beyond which water cannot be raised by a suction pump. If the pump and the piping is tight, however, it will draw water horizontally almost any distance.

Q. Is there any limit in the height to which a pump will force water?

A. None; except the power of the pump.

Q. How do you calculate the power required to pump water?

A. Multiply the number of pounds of water to be pumped per minute by the vertical distance, in feet, between the levels of the supply and discharge, and divide the product by 33,000; the result will be the theoretical horse-power. To this must be added the losses in friction corresponding to the velocity of the water (see page 63). If instead of pumping the water to a higher level it is required to force it against a pressure, multiply by $2\frac{1}{4}$ times the pressure instead of the height, making the same correction for losses as above.

Q. How do you determine the capacity of boiler-feed pumps?

A. Calculate the amount of water which the boiler is capable of evaporating under normal conditions by multiplying the horse-power of the boiler by 30. This will give the number of pounds of water it will evaporate per hour. Divide this by 8.35, which will give the number of gallons. The pump should be capable of supplying about double this quantity, so that it will be adequate when the boiler is forced.

Q. When the water is hot, what precautions must be taken with the pump?

A. It should be brass-lined so that it will not corrode, and it must be placed below the level of the water-supply, as otherwise the hot water will not follow the plunger. It is also advisable to place a valve between the supply and the pump, so that any accumulated vapor may be liberated.

Q. What is an injector?

A. It is an apparatus for forcing water against a pressure by the direct action of a jet of steam upon a mass of water.

Q. Briefly describe the injector and its action.

A. It consists of a steam nozzle through which enters the steam used; a water-supply tube through which enters the water to be forced; a combining tube which begins at the end of the steam nozzle, being that part of the apparatus where the steam and water first come in contact; and, finally, a delivery tube from which the mixture of steam and water enters the discharge pipe. All of these parts have peculiar shapes, which have been determined by years of experimenting; the object being to give the steam and water the proper velocities at different stages in the process. The action of the apparatus may be explained as

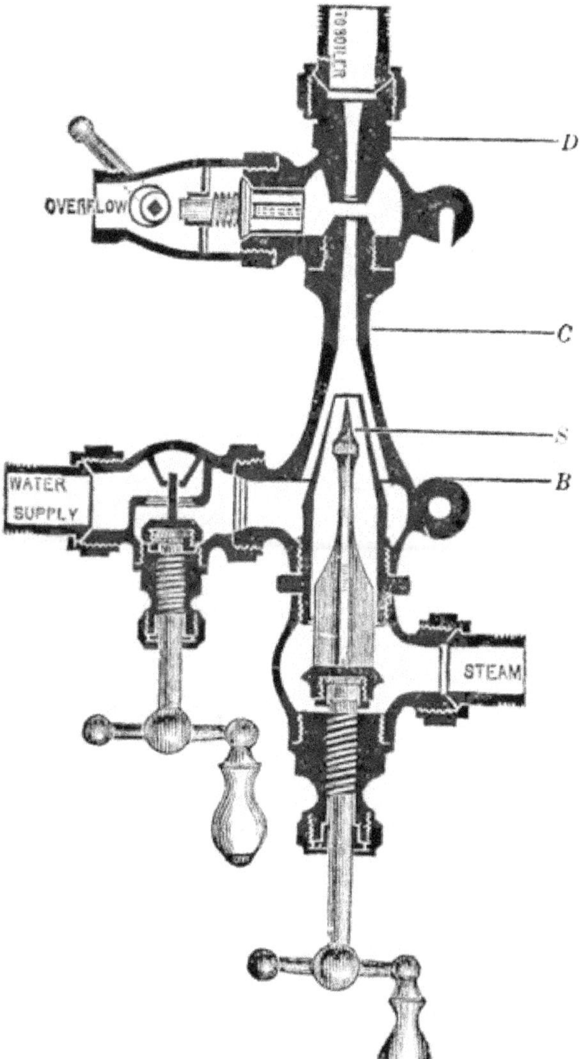

INJECTOR.

S, Steam nozzle.
B, Spinale for adjusting supply of steam.
C, Combining tube.
D, Delivery tube.

follows: The steam leaves the nozzle and enters the combining tube at a high velocity. The friction between the steam jet and the air in the water-supply pipe causes the latter to be exhausted and consequently the water being relieved of the pressure upon its surface soon rises and enters the combining tube, where it comes in contact with the steam jet and condenses it. In being condensed the cross-section of the steam jet is greatly reduced, and the entire energy of its velocity is concentrated upon a very thin jet. This energy being more than sufficient to force it into the boiler, some of it is imparted to the water which it meets in the combining tube, and the entire mixture of steam and water is carried into the delivery tube and thence into the boiler by virtue of the momentum which it has acquired. Of course, the apparatus must be carefully proportioned, since if there is too much water the energy of the condensed steam will not be sufficient to carry it into the boiler, while if there is too little, the steam will not be condensed.

Q. What are the advantages of injectors over pumps?

A. The principal advantages are that water enters the boiler in a steady stream; practically none of the energy of the steam used to operate it is wasted, as all the energy in excess of that necessary to force the water into the boiler is utilized in raising its temperature; the water does not enter the boiler cold—it is more compact and has no moving parts.

Q. What is the commonest cause of the failure of injectors to operate?

A. The presence of air in the suction pipe. This must be avoided by properly packing the valve stem and by entirely submerging the end of the suction pipe. Sediment or dirt in the nozzles will also interfere with the proper working of the apparatus. They should be carefully cleaned out if this occurs.

Q. If the injector does not get water, where would you look for the trouble?

A. It would probably be due to one of the following causes: a leak in the supply pipe, clogging up of the strainer, too hot water, too low a steam pressure for the required lift, or the water-supply may be cut off. I should examine the water pipe first to see that it was intact.

Q. If the injector starts, but afterward the jet breaks, where would you expect to find the difficulty?

A. Any of the causes given in the preceding answer might produce this result, or the trouble might be caused by a loose disc in the valve in the supply pipe, causing it to partly close. In the latter case, the trouble could be remedied by reversing the valve.

Q. What is the difference between lifting and non-lifting injectors?

A. In the former there is a partial vacuum formed in the feed pipe on starting, in the latter a pressure is required in the water-supply.

Q. What are the principal points to be observed in setting up injectors?

A. All pipes, whether steam, water-supply, or delivery, must be of the same or greater internal diameter than the hole in the corresponding branch of each injector, and as short and straight as practicable. When floating particles of wood or other matter are liable to be in the supply water, a strainer must be placed over the receiving end of the water-supply pipe. The holes in this strainer must be as small as the smallest opening in the delivery tube, and the total area of all the holes must be much greater than the area of the water-supply pipe, to compensate for the closing of some of them by deposits. The steam should be taken from the highest part of the boiler, to avoid the carrying over of water with the steam. "Dry pipes" should always be used on locomotives to insure dry steam; wet steam cuts and grooves the steam spindle and steam nozzle. The steam should not be taken from the steam pipe leading to an engine, unless such pipe is large. Sudden variations in pressure may break the jet. After all the pipes are properly connected to the injector and to the boiler, and before steam and water are admitted through them to the injector, they should be disconnected and well washed out by blowing steam or running water through them, to wash out all red lead, scale, or other solids that may be in the pipes. Finally, in setting injectors it is important to place them as low as possible, since their capacity is reduced and the promptness and reliability of their action diminished as the height of lift is increased.

Q. What is an inspirator?

A. It is a double-jet injector—that is, one containing two sets of jets, of which one is used for lifting the water from the source of supply and the other for forcing it into the boiler.

Q. What is an ejector?

A. It is an instrument similar to the injector, but designed for lifting water only, without forcing it against a pressure.

Q. Is an injector more economical than a pump as a boiler feeder?

A. Not always; the injector is the more economical of the two when the feed-water is cold, but the pump is the more economical when the feed-water has been heated.

TABLE*

SHOWING THE RELATIVE EFFICIENCIES OF PUMPS AND INJECTORS.

METHOD OF SUPPLYING FEED-WATER TO BOILER. Temperature of feed-water as delivered to the pump or to the injector, 60° Fahr. Rate of evaporation of boiler, 10 pounds of water per pound of coal from and at 212° Fahr.	Relative amount of coal required per unit of time, the amount for a direct-acting pump, feeding water at 60°, without a heater, being taken as unity.	Saving of fuel over the amount required when the boiler is fed by a direct-acting pump without heater.
Direct-acting pump, feeding water at 60°, without a heater, ...	1.000	.0
Injector feeding water at 150°, without a heater, ...	.985	1.5 per ct.
Injector feeding through a heater in which the water is heated from 150 to 200°,	.938	6.2 "
Direct-acting pump feeding water through a heater, in which it is heated from 60 to 200°, ...	.879	12.1 "
Geared pump, run from the engine, feeding water through a heater, in which it is heated from 60 to 200°,	.868	13.2 "

*Computed by Professor D. S. Jacobus.

Q. Should a boiler plant have both a pump and an injector?

A. Yes, whenever possible; because either the one or the other may at some time refuse to operate. In some cases it would be better to have two pumps, and in others two injectors. (See table on opposite page.)

Q. With what kind of boilers are injectors used the most? Why?

A. With locomotives, because they use cold water, and therefore an injector is more efficient; also because the jarring motion of the engine does not affect an injector in the least, while its effect on the pump would be detrimental. An injector is also much lighter than a pump.

HEATING FEED-WATER.

Q. Why should the feed-water be heated before it enters the boiler?

A. Because cold water fed into a boiler under steam produces strains that will shorten the life of the boiler; because a large proportion of the solid matter frequently contained in water will separate out at a high temperature, and, consequently, if the feed-water is heated sufficiently solids will be deposited in the heater that would otherwise produce scale in the boiler; and because by using exhaust steam, or some other source of heat which

PERCENTAGE OF SAVING IN FUEL BY HEATING FEED-WATER.

STEAM AT 70 POUNDS GAUGE PRESSURE.

TEMPERATURE TO WHICH FEED IS HEATED.

Initial Temperature, Feed	100°	110°	120°	130°	140°	150°	160°	170°	180°	190°	200°	210°	220°	250°	300°
35°	5.53	6.38	7.24	8.09	8.95	9.89	10.66	11.52	12.38	13.24	14.09	14.95	15.81	19.40	29.34
40°	5.12	5.97	6.84	7.69	8.56	9.42	10.28	11.14	12.00	12.87	13.73	14.59	15.45	18.89	28.78
45°	4.71	5.57	6.44	7.30	8.16	9.03	9.90	10.76	11.62	12.49	13.36	14.22	15.09	18.37	28.22
50°	4.30	5.16	6.03	6.89	7.76	8.64	9.51	10.38	11.24	12.11	12.98	13.85	14.72	17.87	27.67
55°	3.89	4.75	5.63	6.49	7.37	8.24	9.11	9.99	10.85	11.73	12.60	13.48	14.35	17.38	27.12
60°	3.47	4.34	5.21	6.08	6.96	7.84	8.72	9.60	10.47	11.34	12.22	13.10	13.98	16.86	26.56
65°	3.05	3.92	4.80	5.67	6.56	7.44	8.32	9.20	10.08	10.96	11.84	12.72	13.60	16.35	26.02
70°	2.62	3.50	4.38	5.26	6.15	7.03	7.92	8.80	9.68	10.57	11.45	12.34	13.22	15.81	25.47
75°	2.19	3.07	3.96	4.84	5.73	6.62	7.51	8.40	9.28	10.17	11.06	12.95	12.84	15.33	24.92
80°	1.76	2.65	3.54	4.42	5.32	6.21	7.11	8.00	8.88	9.78	10.67	11.57	12.46	14.82	24.37
85°	1.30	2.22	3.11	4.00	4.90	5.80	6.70	7.59	8.48	9.38	10.28	11.18	12.07	14.32	23.82
90°	0.89	1.78	2.68	3.58	4.48	5.38	6.28	7.18	8.07	8.98	9.88	10.78	11.68	13.81	23.27
95°	0.45	1.34	2.25	3.15	4.05	4.96	5.86	6.77	7.66	8.57	9.47	10.38	11.29	13.31	22.73
100°	0.00	0.90	1.81	2.71	3.62	4.53	5.44	6.35	7.25	8.16	9.07	9.98	10.88	12.80	22.18

would otherwise be wasted, a very material economy is effected in the consumption of fuel.

A pound of feed-water entering a steam boiler at a temperature of 50° Fahr., and evaporating into steam of 60 pounds pressure, requires as much heat as would raise 1157 pounds of water 1 degree. A pound of feed-water raised from 50° Fahr. to 220° Fahr. requires 170 units of heat; which, if absorbed from exhaust-steam passing through a heater, would be a saving of 15 per cent. in fuel. Feed-water at a temperature of 200° Fahr., entering a boiler, as compared in point of economy with feed-water at 50° Fahr., would effect a saving of over 13 per cent. in fuel; and with a well-constructed heater there ought to be no trouble in raising the feed-water to a temperature of 212° Fahr.

Q. What is the difference between open and closed feed-water heaters?

A. In closed heaters the exhaust steam passes through a series of brass tubes and the water is pumped through the space around the tubes into the boiler, or the water may be pumped through the tubes and the steam pass around the tubes. In the open type, the steam comes in actual contact with the water, the latter passing over a series of cast-iron or steel pans placed in a chamber through which the exhaust steam passes.

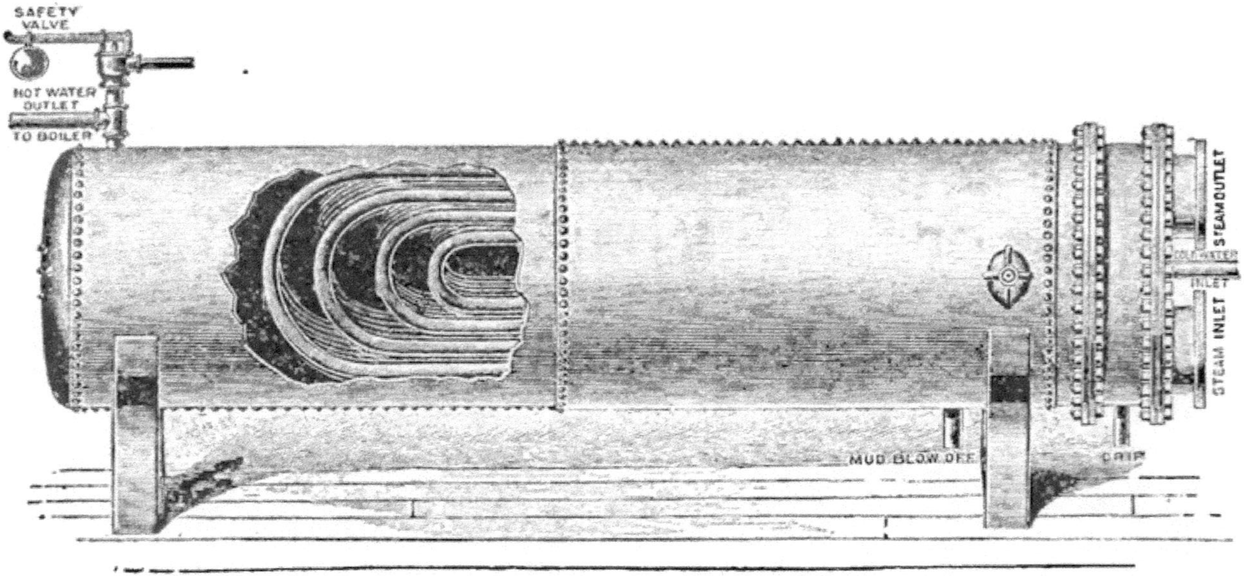

CLOSED HEATER,—BERRYMAN TYPE.

(Steam enters at the inlet pipe shown, and passes through the bent tubes to the steam outlet. Water passes through the space around the tubes, entering at the right and leaving at the upper left. These heaters may be horizontal or vertical, supported on the floor or suspended from ceiling. Tubes of drawn brass; shell, cast iron.)

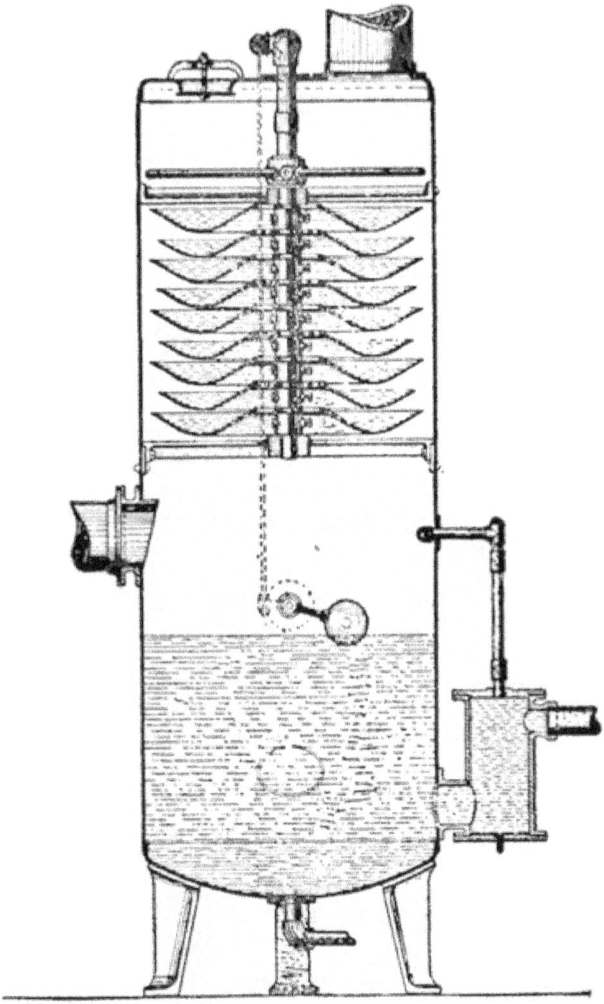

OPEN HEATER,—PITTSBURGH TYPE.

(Steam enters below the pans at the left and passes out at the top. Water enters through the pipe at the top, the flow being regulated by a cock which is controlled by the float and rod. The small cylinder at the right separates the oil. [See also page 172.] The connection to the pump is near the top of the small cylinder. Through an opening in the side of the shell the pans, which rotate around a central shaft, may be cleaned. Shell and pans of steel.)

Q. What is the difference in the method of installing open and closed heaters?

A. In open heaters the pump is placed between the heater and the boiler, hence the pump takes hot water and must therefore be placed below the level of the water in the heater, otherwise the water will not follow the plunger. With the closed type the water enters the pump cold and is forced through the heater into the boiler.

Q. Why can open heaters not be used with injectors?

A. Because if the water is heated to a high temperature, as it should be, in the heater, the injector will not work, it requiring moderately cold water to condense the steam in the combining tube. If the steam in an injector is not condensed the apparatus will refuse to force the water into the boiler.

Q. Which type is, in general, preferable—the open or the closed?

A. Each has its advantages and disadvantages. The closed heater may be located in any convenient position relative to the pump, while the open type must be placed at a higher level than the pump, which, as already stated, has to pump hot water; the open type is not under pressure (except that of the exhaust steam), hence it is lighter and cheaper. It is more easily cleaned; it heats the water to a higher temperature; its purifying properties are better, and it produces no back pressure on the engine. On the other hand, the feed-water may contain grease which will injure the boiler, although it is claimed that by a suitable oil separator this may be entirely eliminated.

Q. What is an economizer?

A. It is a device used for heating the feed-water by means of the products of combustion of the boiler furnace as they pass into the stack.

Q. How is it constructed?

A. The economizer usually consists of a series of cast-iron or steel tubes connected at either end by headers similar to those used in water-tube boilers. The water circulates through the tubes, which are placed in the flue connection just at the entrance to the stack.

Q. What fittings should an economizer have?

A. As it is virtually a water-tube boiler, it should have a blow-off pipe and a safety-valve, because if the boiler is not supplying steam as usual the water in the economizer tubes will be evaporated, producing an excessive pressure.

Q. For what purpose are economizers generally used?

A. For the purpose of increasing the capacity or efficiency of existing boiler plants.

Q. Why are they generally not necessary in new installations?

A. Because if the boilers are properly constructed they do not allow much heat to be wasted through the chimney.

Q. What other method of heating the feed-water is sometimes used?

A. It is heated by the use of condensers in connection with the engines. (See "Condensers," page 233.)

FURNACES AND FLUES.

Q. Can you calculate the strength of a flue by the same rules that apply to the shells of boilers?

A. No; because the same rules for strength of cylinders under pressure from within do not apply to those which are subjected to a pressure from without.

Q. If pressure is exerted on the internal or external surface of the cylinder, is the effect not the same in both cases?

A. No; when pressure is exerted within a tube or cylinder, the tendency of the strain is to cause the tube to assume the true cylindrical form; but when pressure is exerted on the outside of the tube, the tendency of that pressure is to crush the tube or flatten it; as it is a well-known fact that iron of any strength when formed into a tube will require a much greater strain to tear it asunder than it would take to crush it. A thin hoop of iron will resist a very great amount of tearing force, but if that same hoop or circle be placed as a prop under half the weight that was exerted to tear it apart, it would be crushed flat.

Q. What is the difference between external and internal strain?

A. Internal is a tearing strain, while external is a crushing strain; and flues and tubes of boilers are nothing but a series of props, and a constant tendency of the pressure is to flatten the tube or flue and cause it to collapse.

Q. What is a collapse?

A. It is the crushing or flattening of a flue by overpressure, and is often attended with terrible results.

Q. How do you calculate the strength of flues or cylinders subjected to external pressure?

A. It has been shown by experiment that the strength of such cylinders is proportional to the square of the thickness of the cylinder and inversely proportional to the length and to the diameter. The formula for collapsing is:

$$P = 806{,}000 \frac{t^2}{l\,d},$$

where P is the collapsing pressure in pounds per square inch,

l is the length of the cylinder in feet,

d is the diameter of the cylinder in inches.

Rule for Finding the Collapsing Pressure of a Cylindrical Flue.—Multiply the square of the thickness in inches by the number 80,600. Multiply the length of the flue in feet by its diameter in inches. Divide the first product by the second, and the quotient will be the collapsing pressure in pounds per square inch.

Q. If the length of a cylindrical flue is 10 feet, its diameter 2 feet, and thickness ¼ inch, what will be the collapsing pressure?

A. $P = \frac{806{,}000 \times \frac{1}{4} \times \frac{1}{4}}{10 \times 24} = 215$ pounds.

Q. How may long flues be strengthened?

A. This may be done in various ways. The old method was to rivet rings of angle- or tee-iron around the flue at fixed intervals, or to make the flue in sections and to join them together by riveting on ∩-shaped rings. The modern method is to make the entire flue of corrugated iron, which not only adds strength, but facilitates expansion and increases the heating surface.

Q. When the flue is stiffened by rings, as described above, how do you calculate its strength?

A. By the same rule as that for plain flues, except that the length between rings is taken as the length of the flue.

Q. What method is employed in the Galloway boiler for strengthening the flues?

A. The Galloway tubes, which are conical in form and placed within and across the flues, being riveted to the sides.

GRATES.

Q. What is the simplest form of grate?

A. It consists of a series of cast-iron bars shaped like beams, supported at either end, and so placed as to allow spaces between them for the passage of air.

Q. What points should, in general, be observed in grates?

A. They should be flat on top and supported, but not fixed at the ends, as otherwise the expansion and contraction will cause them to get out of shape. The spaces between the bars should be numerous and as large as possible. The width of the spaces, however, depends on the kind of coal to be used, and in practice varies from $\frac{3}{8}$ to $\frac{5}{8}$ inch. The height of the grate above the bottom of the ashpit should be from 24 to 30 inches, and the bars should, in general, be inclined downward toward the bridge wall, as the fuel may then be more easily distributed. The length is limited by the distance to which a fireman can throw the coal, which is about 6 feet.

Q. How much coal is generally consumed per square foot of grate surface?

A. This depends on the nature of the draught and the kind of coal. For land boilers fired with a good quality of anthracite coal, 9 pounds per square foot is a fair average. In some boilers operating under a light draught the coal consumption is as low as 4 pounds, while in locomotives using a blast pipe to produce a strong draught as high as 120 pounds of coal may be burned per square foot of grate surface per hour

Q. How much grate surface should be allowed per horse-power?

A. In land boilers about ⅓ square foot of grate surface is given per horse-power. With good bituminous coal, better results are obtained by using a smaller grate area and a strong draught. With coal containing a high percentage of ash it is better to use a large grate surface with a comparatively slow rate of combustion.

Q. What is a shaking grate?

A. It is a grate designed for cleaning the fire breaking up clinkers, and removing them without opening the fire door.

Q. What are the advantages to be derived from such an arrangement?

A. Whenever the fire doors are opened cold air rushes in, tending not only to impair the efficiency of the boiler, but also its durability. Moreover, it is impossible for a fireman to thoroughly stir out, with a slicing-bar, every part of the grate. Hence, if the coal has a tendency to form clinkers the advantages of a shaking grate would be material.

Q. What is meant by automatic stoking?

A. A system by which the coal is fed to, and the ashes removed from, the furnace automatically without opening the furnace doors.

Q. How long have automatic or mechanical stoking devices been in use?

A. A device similar in many respects to the modern mechanical stokers was employed by Watt in 1785.

Q. Under what conditions are mechanical stokers especially desirable?

A. When the fuel used consists of mine refuse, screenings, or other materials not generally used in manual firing.

Q. What advantages are claimed for mechanical stokers?

A. Fuel economy, prevention of smoke, saving in labor, and cleanliness in the boiler room.

Q. Why is mechanical stoking productive of economy in the use of fuel?

A. Because the coal is spread upon the grate uniformly and at the time when it is needed. With hand-firing the coal is fed to the furnace at irregular intervals, and usually more coal is put on than necessary. Besides, each time the boiler is fired and cleaned, the furnace doors are opened and cold air rushes in. All of these features which attend hand-firing are injurious to the economy of operation. With a system of mechanical stoking they are not incurred, and hence the efficiency may be materially increased.

Q. Why do mechanical stokers lessen the production of smoke?

A. Because the fuel is fed uniformly in small quantities instead of intermittently and in bulk, as in the case of hand-firing. A uniform temperature is maintained in the furnace, and the motion of the grate keeps the spaces open for the continual passage of the air. Hence the combustion is at all times complete, which means absence of smoke.

Q. Why are they productive of saving in labor?

A. Because there is no cleaning of fires or manual labor of any kind, except, perhaps, the bringing of the coal to the hoppers; and even this is frequently accomplished by machinery.

Q. Why are they more cleanly?

A. Because the usual dirty appearance of boiler plants is produced by the dust raised in shoveling the coal, cleaning the fires, and removing the ashes, all of which operations are abolished in mechanical stoking.

Q. Do mechanical stokers pay in small plants?

A. No, they do not; because the cost of the plant and the power consumed in operating would not be warranted by the saving which would accrue.

CHIMNEYS AND STACKS.

Q. What is the object of a chimney or stack?

A. It is for the purpose of producing a draught, ejecting the products of combustion, and supplying fresh air for the combustion of the fuel.

Q. How does a chimney produce a draught?

A. The tendency of the rarefied gases is to rise, producing a partial vacuum which causes a rush of air through the furnace.

Q. Which kinds of coal require the tallest stacks?

A. Anthracites, because they do not burn as readily as bituminous coals.

Q. On what does the draught produced by a chimney depend?

A. It depends on two factors: on the height of the chimney and on the difference in weight of the gases contained in the chimney and the atmosphere.

Q. On what does this difference in weight largely depend?

A. Upon the temperature of the gases leaving the boiler.

Q. At what temperature do the gases usually leave in well-designed boilers?

A. 500 to 600 degrees Fahrenheit.

Q. At what temperature of the escaping gases is the best draught obtained?

A. At about 580 degrees Fahrenheit.

Q. On what does the area of the chimney for a given boiler plant depend?

A. It depends upon the quantity of coal consumed.

Q. What relation is there between the quantity of coal consumed and the area of the chimney?

A. The area of the cross-section in square inches should be from $1\frac{1}{2}$ to 2 times the number of pounds of coal consumed per hour.

Q. According to this rule, what would be the proper diameter of chimney for 500 horse-power boilers of the water-tube type?

A. Assuming an evaporation of 10 pounds of water under normal conditions per pound of coal, we have:

Pounds of water evaporated per pound of coal $= 10$.

Total pounds of water evaporated per hour $= 30 \times 500 = 15{,}000$.

Pounds of coal consumed per hour

$$= \frac{15{,}000}{10} = 1500.$$

Area of chimney $= 1500 \times 1\frac{1}{2}$ to 1500×2
$= 2250$ to 3000 square inches.

Diameter of chimney $= 53\frac{1}{2}$ to $61\frac{3}{4}$ inches or, say, 60 inches.

Q. What is the relation between grate and chimney area?

A. A fair average of coal consumed per square foot of grate surface for anthracite coal is 12 pounds. Hence the chimney area being about $1\frac{3}{4}$ square inches per pound of coal, we have:

Chimney area per pound of coal $= 1\frac{3}{4}$ square inches.

Chimney area per square foot of grate surface $= 1\frac{3}{4} \times 12 = 21$ square inches $= \frac{21}{144}$ $= \frac{1}{7}$ square foot;

or, in other words, the chimney area should be about $\frac{1}{7}$ of the grate area.

Q. Is there any relation between the cross-section of chimney and horse-power?

A. For fire-tube boilers the average heating surface is 12 square feet per horse-power, while the ratio of grate to heating surface is about 1 : 35. Hence the grate surface per horse-power may be taken roughly as $\frac{12}{35}$, or about $\frac{1}{3}$. If, now, we take the results above, we have for the chimney area per horse-power, $\frac{1}{3} \times \frac{1}{7} = \frac{1}{21}$ for fire-tube boilers, and a trifle smaller, say $\frac{1}{23}$, for water-tube boilers.

Q. What determines the height of chimneys?

TABLE FOR CHIMNEY DIMENSIONS.*

Diameter in Inches.	Height of Chimneys. 50 ft.	60 ft.	70 ft.	80 ft.	90 ft.	100 ft.	110 ft.	125 ft.	150 ft.	175 ft.	200 ft.	Effective Area, Square Feet.	Actual Area, Square Feet.	Side of Square of Approximate Area, Inches.
	Commercial Horse-Power.													
18	23	25	27									0.97	1.77	16
21	35	38	41									1.47	2.41	19
24	49	54	58	62								2.08	3.14	22
27	65	72	78	83								2.78	3.98	24
30	84	92	100	107	113							3.58	4.91	27
33		115	125	133	141							4.47	5.94	30
36		141	152	163	173	182						5.47	7.07	32
39			183	196	208	219						6.57	8.30	35
42			216	231	245	258	271					7.76	9.62	38
48				311	330	348	365	389				10.44	12.57	43
54				363	427	449	472	503	551			13.51	15.90	48
60				505	539	565	593	632	692	748		16.98	19.64	54
66					658	694	728	776	849	918	981	20.83	23.76	59
72					792	835	876	934	1023	1105	1181	25.08	28.27	64
78						995	1038	1107	1212	1310	1400	29.73	33.18	70
84						1163	1214	1294	1418	1531	1637	34.76	38.48	75
90						1344	1415	1496	1639	1770	1893	40.19	44.18	80
96						1537	1616	1720	1876	2027	2167	46.01	50.27	86

* From "Steam,"—Babcock and Wilcox, 1896.

A. The height of chimneys is determined by the required draught. It is influenced by the kind of coal to be burned as well as by its location, as it must, in general, be higher than hills or buildings in the immediate vicinity.

STEAM SEPARATORS AND TRAPS.

Q. For what purpose are steam separators used?

A. For removing moisture from steam before it enters the engine cylinder; or they may be used for extracting other liquids from vapors, as, for example, the oil contained in exhaust steam. The first named is generally called a live steam separator.

Q. Why should it be advisable to extract the entrained water from steam before using it in the engine?

A. Because an accumulation of water in the cylinder is often the cause of blowing out the head of the cylinder or steam-chest cover; and also because the presence of moisture in steam reduces the economy of the engine.

Q. How should a separator be constructed to be efficient?

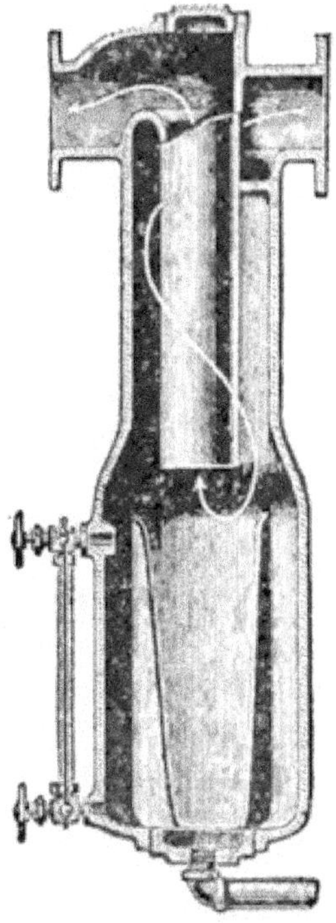

STEAM SEPARATOR.
STRATTON TYPE.

A. The steam entering the apparatus at a high velocity should have its direction of flow altered or reversed, so as to destroy the momentum of the liquid particles, permitting them to fall by gravity into a vessel provided for that purpose. This being accomplished, the steam should not again come in contact with the water, as it is liable to pick up particles of any liquid with which it comes in contact. Finally, the cross-section for the passage of the steam should be ample in all parts of the apparatus, so that the losses by friction will be reduced to a minimum.

Q. For what other purpose are separators frequently used?

A. To extract the oil from feed-water in open heaters.

Q. How are these constructed?

A. In various ways. In the Pittsburgh heater, illustrated on page 157, the separation of oil is accomplished by means of a small cylinder placed on the side of the apparatus near the bottom. This cylinder is connected by pipes to the steam- and water-spaces of the heater, as shown in the cut; the feed to the pump is at the top of the small cylinder. As the oil floats on the surface of the water it is evident that none will find its way into the small cylinder, so long as the water is maintained at its proper level, while if the level of the water should become too low the pump will not be supplied with water.

Q. For what purpose are steam traps used?

A. For the purpose of removing condensed steam from a system of steam piping, without allowing any of the steam itself to escape.

Q. How is this accomplished?

A. The trap is connected to the piping to be drained and contains an outlet controlled by a valve. The valve in some traps is operated by a float, and in others by means of a bent tube of elliptical cross-section. In the former the opening and closing of the valve is determined directly by the amount of water in the trap. In the curved-tube system the opening and closing of the valve depend upon the temperature.

Q. Suppose a separator, trap, heater, or other appliance should require cleaning or repairing, will it not be necessary to shut down the plant?

A. No; they should always be provided with by-passes for both steam and water, that is, they should be connected with the piping in such a way that the steam or water may be made to pass temporarily through auxiliary pipes around the heater trap or other appliance.

Q. Give a brief description of the manner in which a by-pass is usually constructed.

A. As generally constructed a by-pass consists

of a pipe leading around the appliance and fitted with three valves—V, V,, and V,,,—as shown in the accompanying cut, the trap (in this case) being connected to the piping by pipe unions U, U. Under ordinary conditions, that is, when the trap is in operation, the valves V, and V,, remain open while V,,, is closed. If, however, the trap is to be taken out for any reason, it is only neces-

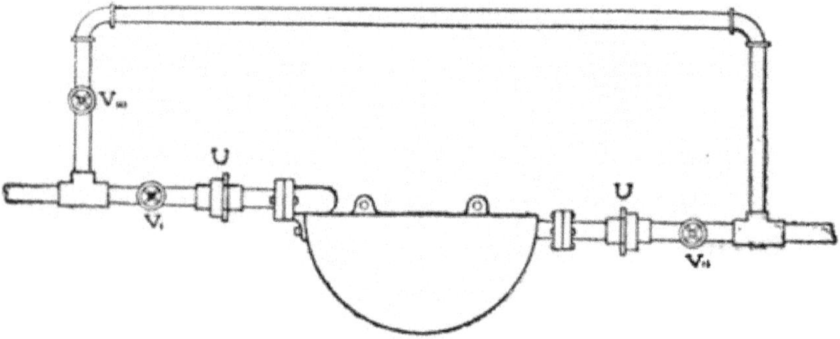

sary to close the valves V, and V,, and to open V,,,. The steam, instead of passing through the trap, will then pass around it through the by-pass, and the trap or other appliance may be disconnected by means of the two unions U, U, without in any way interfering with the operation of the plant. For feed-water heaters, etc., a similar by-pass should be provided for the water.

THE STEAM ENGINE.

The steam engine, as it exists to-day, may be said to be the invention of James Watt. While he was not the originator of the idea of utilizing the pressure and expansive force of steam for the purpose of doing mechanical work, Watt's discoveries and inventions, in this connection, were of such importance that he is generally considered as the inventor of the steam engine.

In looking over the models of engines and accessories of James Watt, a great many of which are exhibited in the South Kensington Museum, London, it is surprising to note how little change the steam engine has undergone during the past century. It is to-day, in fact, the same machine that it was then; and while the results which have since been accomplished in the way of economy, regulation, speed, and power doubtless exceed the most sanguine expectations of the early workers in this field, the modern engine is, nevertheless, practically the same machine that it was a century ago.

The efforts of steam engineers, since the days of James Watt, have produced not only vastly more powerful machines, higher and more uniform speed and what now seems perfect running, but they have also very materially increased the efficiency of the engine. And yet the results which have been obtained in the way of economy still leave much to be desired. The steam engine and boiler, considered as an apparatus for converting the potential energy contained in coal or other fuel into mechanical work, is a most extravagant machine. With the very best engines and boilers we are not able to develop a horse-power with a consumption of much less than 3 pounds of coal per hour, while if all of the energy were utilized we should obtain from that amount of good coal not less than 14 horse-power. In other words, the best engines and boilers utilize only about 7 per cent. of the latent energy cf the fuel. As far as the engine itself is concerned, the mechanism leaves but little to be desired. In such engines as are generally used for electric lighting, that is, the high-speed automatic cut-off type, the regulation is such that the full load may be suddenly thrown on or off without producing a variation in the speed of the engine greater than 1 to 2 per cent., and at all loads such engines, when properly adjusted, run smoothly, noiselessly, and without producing vibration.

HORSE-POWER.

Q. What is meant by the power of a steam engine?

A. The amount of work it will do in a given space of time.

Q. Define the unit of power.

A. The unit generally adopted for the power of steam engines is the *horse-power.* An engine of 1 horse-power means one which will raise 550 pounds 1 foot a second or its equivalent.

Q. What would be equivalent to this amount of work?

A. As work is the product of force times space, a weight of 550 pounds raised 1 foot would be equal to 550 foot-pounds of work. If 1 pound were raised 550 feet or 2 pounds 275 feet, the amount of work would be the same. Hence, a horse-power may be defined as 550 foot-pounds per second, 33,000 foot-pounds per minute, 1,980,-000 foot-pounds per hour, and so on.

Q. Name some form of work other than raising a weight, which would be equivalent to 1 horse-power.

A. An electric current of 10 ampères at 74.6 volts.

Q. What determines the horse-power of a steam engine?

A. The diameter of the cylinder, length of stroke, average or mean effective pressure on the piston, and the speed.

Q. How do you calculate the horse-power of an engine?

A. By multiplying the area of the piston in square inches by the mean effective pressure acting upon it; multiplying the length of stroke in feet by the number of strokes (twice the number of revolutions) per minute; multiplying the first product by the second, and dividing by 33,000.

Q. What would be the horse-power of an 18″ x 18″ engine at 200 revolutions per minute, with a mean effective pressure of 45 pounds per sq. inch?

A. Area of piston $= 18 \times 18 \times .7854 = 254$ square inches,

Total mean pressure on piston $= 254 \times 45 = 11,430$ pounds,

Number of strokes per minute $= 2 \times 200 = 400$,

Length of stroke $= 18 \div 12 = 1.5$ feet,

Distance traveled by piston per minute $= 400 \times 1.5 = 600$ feet,

Work done per minute $= 11,430 \times 600 = 6,858,000$ foot-pounds,

Horse-power $= 6,858,000 \div 33,000 = 208$.

Q. How would you write the above rule in the shape of a formula?

A. Let HP = horse-power,
P = mean effective pressure in pounds per square inch,
L = length of stroke in feet,
A = area of piston in square inches,
N = number of strokes per minute,
R = number of revolutions per minute,
S = piston speed in feet per minute,
d = diameter of cylinder in inches;

the formula corresponding to the above rule would be:

$$(A = .7854\, d^2)$$

$$HP = \frac{P\, L\, A\, N}{33{,}000} \text{ or } \frac{A\, P\, S}{33{,}000}.$$

Q. Given the horse-power, mean effective pressure, and piston speed, how would you find the proper diameter of cylinder? Give rule and formula.

A. The formula would be

$$d = \sqrt{\frac{42{,}017\, HP}{P\, S}} \text{ or } 205\sqrt{\frac{HP}{P\, S}}$$

and the rule as follows: Multiply the horse-power by 42,017; multiply the piston speed by the mean effective pressure; divide the first product by the second and extract the square root of the quotient.

Q. Write formulæ for length of stroke, piston speed, and number of revolutions when the other quantities are given.

A. Length of stroke $= L = \dfrac{33{,}000\ HP}{P\ A\ N} = \dfrac{16{,}500\ HP}{P\ A\ R} = \dfrac{21{,}010\ HP}{P\ R\ d^2}$;

Piston speed $= S = N\ L = 2\ R\ L = \dfrac{33{,}000\ HP}{P\ A}$;

Number of revolutions $= R = \dfrac{S}{2\ L} = \dfrac{16{,}500\ HP}{P\ A\ L} = \dfrac{21{,}010\ HP}{P\ L\ d^2}$

Q. What do you understand by the mean effective pressure?

A. The average forward pressure on the piston less the back pressure.

Q. What is the average forward pressure?

A. It is a pressure depending upon the initial pressure in the cylinder and the point of cut-off.

Q. How do you find the average (forward) pressure in a given case?

A. In the following table look up the multiplier corresponding to the cut-off. To the initial gauge pressure in the cylinder add 14.7 pounds to obtain the initial absolute pressure. Multiply this by the number corresponding to the cut-off in the table, and the product will be the absolute average forward pressure.

Q. What would be the average pressure corresponding to 80 pounds initial by the gauge and $\frac{1}{4}$ cut-off?

A. $80 + 14.7 = 94.7 \times .5965 = 56.45$ absolute

$$\begin{array}{r} 56.45 \\ 14.7 \\ \hline 41.75 \end{array}$$ gauge.

TABLE

OF MULTIPLIERS FOR MEAN ABSOLUTE PRESSURES.

CUT-OFF.	RATE OF EXPANSION.	MULTIPLIER.	CUT-OFF.	RATE OF EXPANSION.	MULTIPLIER.
$\frac{1}{4}$	4	.5965	$\frac{5}{8}$	1.6	.9188
$\frac{1}{3}$	3	.6995	$\frac{2}{3}$	1.5	.9370
$\frac{3}{8}$	2.66	.7428	$\frac{3}{4}$	1.33	.9657
$\frac{1}{2}$	2	.8465	$\frac{7}{8}$	1.14	9919

Q. How do you find the mean effective pressure?

A. Find the absolute mean forward pressure as described above and deduct the absolute back pressure.

Q. What is the back pressure?

A. It is the pressure opposing the piston. In engines exhausting into the atmosphere it is usually about 15 pounds per square inch (atmospheric pressure). In condensing engines it varies from two (2) pounds per square inch up to atmospheric pressure, depending on the vacuum. Where the exhaust is used in a heating system, it varies from 16 to 25 pounds, depending on the amount of friction in the piping.

Q. What horse-power would be developed by an engine under the following conditions:

Stroke, 12 inches;

Diameter of cylinder, 12 inches;

Initial gauge pressure, 80 pounds per square inch;

Speed, 300 revolutions per minute;

Back pressure (gauge), 5 pounds per square inch;

Cut-off, $\frac{1}{4}$.

A. The absolute initial pressure is $80 + 14.7 = 94.7$ pounds, and the multiplier in the table corresponding to $\frac{1}{4}$ cut-off being .5965, the average forward pressure is $94.7 \times .5965 = 56.45$ pounds absolute. The back pressure being $5 + 14.7 = 19.7$ pounds absolute, the mean effective pressure is $56.45 - 19.7 = 36.75$ pounds per square inch.

Area of piston $= 12 \times 12 \times .7854 = 113.1$ square inches.

Total mean pressure on piston $= 36.75 \times 113.1 = 4153$ pounds.

Length of stroke = 12 ÷ 12 = 1 foot.

Number of strokes = 300 × 2 = 600 per minute.

Distance traveled by piston = 600 × 1 = 600 feet per minute.

Work done per minute = 600 × 4153 = 2,491,800.

Horse-power = 2,491,800 ÷ 33,000 = 75 horse-power.

Q. If in the above example, instead of exhausting against a back pressure, a condenser had been used, in which there was a vacuum of 22 inches, what would have been the gain in power?

A. Since each inch of vacuum corresponds to about ½ pound, the back pressure would be 22 × ½ = 11 pounds less than atmospheric, or 14.7 — 11 = 3.7 pounds absolute. Hence the mean effective pressure = 56.45 — 3.7 = 52.75 pounds, and the horse-power $\frac{52.75 \times 113.1 \times 600}{33,000} = 108.$ That is, the gain in power would be 108 — 75 = 33 horse-power, or over 40 per cent.

EXPLANATION OF TABLE.

The table on the following pages is calculated for different cylinder diameters from 4 inches to 5 feet and for piston speeds of 300 to 600 feet per minute. To find the horse-power of any engine

TABLE

OF HORSE-POWER FOR DIFFERENT CYLINDER DIAMETERS AND PISTON SPEEDS.

HORSE-POWER PER POUND MEAN EFFECTIVE PRESSURE.

Diameter of Cylinder.	Speed of Piston in Feet per Minute.						
	300	350	400	450	500	550	600
Inches.							
4	.114	.133	.152	.171	.19	.209	.228
4½	.144	.168	.192	.216	.24	.264	.288
5	.18	.21	.24	.27	.30	.33	.36
5½	.216	.252	.288	.324	.36	.396	.432
6	.256	.299	.342	.385	.428	.471	.513
6½	.307	.391	.409	.461	.512	.563	.614
7	.348	.408	.466	.524	.583	.641	.699
7½	.401	.468	.534	.602	.669	.735	.802
8	.456	.532	.608	.685	.761	.837	.912
8½	.516	.602	.688	.774	.86	.946	1.032
9	.577	.674	.770	.866	.963	1.059	1.154
9½	.644	.751	.859	.966	1.074	1.181	1.288
10	.714	.833	.952	1.071	1.390	1.309	1.428
10½	.787	.919	1.050	1.181	1.313	1.444	1.575
11	.864	1.008	1.152	1.296	1.44	1.584	1.728
11½	.943	1.1	1.257	1.414	1.572	1.729	1.886
12	1.025	1.195	1.366	1.540	1.708	1.880	2.050
13	1.206	1.407	1.608	1.809	2.01	2.211	2.412
14	1.398	1.631	1.864	2.097	2.331	2.564	2.797
15	1.606	1.873	2.131	2.409	2.677	2.945	3.212
16	1.827	2.131	2.436	2.741	3.045	3.349	3.654
17	2.054	2.396	2.739	3.081	3.424	3.766	4.108
18	2.312	2.697	3.083	3.468	3.854	4.239	4.624
19	2.577	3.006	3.436	3.865	4.295	4.724	5.154
20	2.855	3.331	3.807	4.265	4.759	5.234	5.731
21	3.148	3.672	4.197	4.722	5.247	5.771	6.296
22	3.455	4.031	4.607	5.183	5.759	6.334	6.911
23	3.776	4.405	5.035	5.664	6.294	6.923	7.552
24	4.111	4.797	5.482	6.167	6.853	7.538	8.223
25	4.461	5.105	5.948	6.692	7.436	8.179	8.923
26	4.826	5.630	6.435	7.239	8.044	8.848	9.652
27	5.199	6.066	6.932	7.799	8.666	9.532	10.399
28	5.596	6.529	7.462	8.395	9.328	10.261	11.193
29	6.006	7.007	8.008	9.009	10.01	11.011	12.012

HORSE-POWER PER POUND MEAN EFFECTIVE PRESSURE.

Diameter of Cylinder.	Speed of Piston in Feet per Minute.						
	300	350	400	450	500	550	600
Inches.							
30	6.426	7.497	8.568	9.639	10.71	11.781	12.852
31	6.865	8.001	9.144	10.287	11.43	12.573	13.716
32	7.308	8.526	9.744	10.962	12.18	13.398	14.616
33	7.770	9.065	10.360	11.655	12.959	14.245	15.54
34	8.238	9.611	10.984	12.357	13.73	15.103	16.476
35	8.742	10.199	11.656	13.113	14.57	16.027	17.484
36	9.252	10.794	12.336	13.878	15.42	16.962	18.504
37	9.774	11.403	13.032	14.861	16.29	17.919	19.548
38	10.308	12.026	13.744	15.462	17.18	18.898	20.616
39	10.86	12.67	14.48	16.29	18.1	19.91	21.62
40	11.424	13.328	15.232	17.136	19.04	20.944	22.848
41	12.006	14.007	16.008	18.009	20.00	22.011	24.012
42	12.594	14.693	16.792	18.901	20.99	23.089	25.188
43	13.20	15.4	17.6	19.8	22.0	24.2	26.4
44	13.818	16.121	18.424	20.727	23.03	25.333	27.636
45	14.454	16.863	19.272	21.681	24.09	26.339	28.908
46	15.128	17.626	20.144	22.662	25.18	27.698	30.216
47	15.768	18.396	21.024	23.652	26.28	28.908	31.536
48	16.446	19.187	21.928	24.669	27.41	30.151	32.152
49	17.142	19.999	22.856	25.713	28.57	31.427	34.284
50	17.85	20.825	23.8	26.775	29.75	32.725	35.7
51	18.54	21.665	24.76	27.855	30.95	34.045	37.08
52	19.296	22.512	25.728	28.944	32.16	35.376	38.592
53	20.052	23.394	26.736	30.078	33.42	36.762	40.104
54	20.82	24.29	27.76	31.23	34.7	38.17	41.64
55	21.594	25.193	28.792	32.391	35.99	39.589	43.188
56	22.386	26.117	29.848	33.579	37 31	41.041	44.772
57	23.196	27.062	30.928	34.794	38.66	42.526	46.392
58	24.018	28.021	32.024	36.027	40.03	44.033	48.036
59	24.852	28.994	33.136	37.278	41.42	45.562	49.704
60	25.698	29.981	34.264	38.547	42.83	47.113	51.396

by means of this table, multiply twice the number of revolutions per minute by the length of stroke in feet. This will give the piston speed in feet per minute. Look up the horse-power from the table for this piston speed and the proper diameter of cylinder and multiply it by the mean effective pressure. Take the above example as an illustration; the piston speed was found to be 600 feet per minute, and hence for a 12-inch cylinder the horse-power from the table is 2.05 for each pound of mean effective pressure. Hence multiplying this by the mean effective pressure, 52.75, we have $52.75 \times 2.05 = 108$ horse-power.

Q. Is the pressure in the boiler and the pressure in the cylinder nearly equal in all cases?

A. No; the pressure in the cylinder is in many cases less than the pressure in the boiler.

Q. From what causes does the difference between the pressure in the boiler and the pressure in the cylinder arise?

A. First, from a malconstruction of the steam-pipe and steam-ports; *secondly*, from loss by radiation and condensation; *thirdly*, from the action of the governor; and, *fourthly*, from the bad condition of the piston.

Q. What is the most economical steam pressure to use in the cylinder of a high-pressure engine?

A. From 80 to 90 pounds to the square inch.

Q. Why should 80 or 90 pounds to the square inch be more economical than lower pressure, say 40 or 45 pounds to the square inch?

A. On account of the back pressure of the

atmosphere; for instance, if we have a pressure of 45 pounds to the square inch on the piston, the loss by atmospheric pressure is 15 pounds to the square inch, which is about $\frac{1}{3}$ of the pressure on the piston, leaving only 30 pounds for useful effect and to overcome the friction of the engine; if we have a pressure of 90 pounds to the square inch, the loss is only 15 pounds to the square inch, or about $\frac{1}{6}$.

Q. Is it economical to use an engine that is too large for the work to be done?

A. No; because an engine running below its rated load wastes steam. If it is a *throttling engine,* the steam is throttled, or reduced without doing work, which means a loss. If it is an automatic cut-off engine the expansion is increased, which also impairs the economy of the engine.

Q. Why does increasing the rate of expansion reduce the economy?

A. There is one point of cut-off which is more economical than any other, because at that point the steam expands to atmospheric pressure and is not capable of doing any more work when exhausted. This cut-off, for an initial pressure of 80 pounds, is $\frac{1}{7}$. If the rate of expansion is reduced, the steam is exhausted before it has done as much work as it is capable of doing, while if the rate of expansion is increased, the terminal pressure is liable to fall below that of the atmosphere, in which case the opposing pressure of the atmosphere will retard the piston during the latter part of its stroke. This also means a waste of power.

DIFFERENT KINDS OF ENGINES.

Q. What is the difference between condensing and non-condensing engines?

A. In non-condensing engines the steam, after having done its work in the steam cylinder, escapes into the atmosphere, or sometimes into a heating system where the heat still contained in the steam is partially utilized. In the condensing engine the steam exhausts into a condenser, where it comes in contact with some cooling medium, in consequence of which it is condensed, producing a partial vacuum behind the piston.

Q. What is the object of condensing?

A. To increase the effective pressure on the piston and consequently the power.

Q. By how much is the power of a non-condensing engine increased when a condenser is added?

A. The power is increased in the ratio which the vacuum in the condenser bears to the mean effective pressure.

Q. Suppose an engine working at 80 pounds

initial pressure and $\frac{1}{4}$ cut-off exhausting against the atmosphere, had a condenser added. If there were an effective vacuum of 26 inches, what would be the percentage increase in power if the speed remained the same?

A. According to the rules given above, the mean effective pressure was originally

$$(80 + 14.7) \times .5965 - 14.7 = 41.75 \text{ pounds},$$

which was increased by adding a condenser whose vacuum is 26 inches by

$$26 \div 2 = 13 \text{ pounds}.$$

Hence the increase in power is

$$\frac{13}{41.75} = 31 \text{ per cent.}$$

Q. Does it not require power to operate a condenser?

A. Yes; but generally not so much as is gained by its use.

Q. What percentage is gained in economy by condensing?

A. From 20 to 35 per cent., depending on the type and size of engine.

Q. Why, then, are not all engines built for condensing?

A. Because in small engines the saving in fuel would not be enough to warrant the additional first cost, and the increased labor and attention which the plant would require. Further, in many installations the steam leaving at atmospheric pressure can be used to good advantage for heating purposes or for purifying the water before it enters the boiler. Finally, in cities the cost of the water is frequently in excess of what would be saved in fuel.

Q. How much water is required for condensing?

A. About 25 times as much as passes through the engine.

(See also "Condensers," page 233.)

Q. What do you mean by "simple" or single expansion and by multiple expansion engines?

A. A simple or single expansion engine is one in which the steam is used expansively in one cylinder or set of cylinders only, and after exhausting is not used again for doing work in the engine. In multiple expansion engines the steam expands successively, doing work, in two or more cylinders or sets of cylinders.

Q. What are the names given respectively to engines in which the steam expands two, three, and four times?

A. Compound, triple expansion, and quadruple expansion engines.

Q. What is meant by compounding?

A. By the term "compounding" is meant expanding the steam successively in two or more cylinders.

Q. Why are engines compounded?

A. To secure greater economy in the use of steam.

Q. Is not the friction of an engine greater if it uses the same amount of steam in two or three cylinders than if the entire work is performed in a single cylinder?

A. Yes; because each cylinder (except in tandem compound engines) has its own crank and attending mechanism.

Q. Why, then, is expanding successively in several cylinders productive of economy in the use of fuel?

A. The higher the initial steam pressure used in a steam-power plant, and the lower the terminal pressure (provided it is not less than the back pressure), the greater the economy. Hence, in order to secure the greatest fuel economy, there must of necessity be a wide range of temperature from live to exhaust steam. If the expansion occurred in a single cylinder, the walls of the latter and a portion of the steam passages would be subjected to this variation in temperature at each stroke. In other words, the cylinder walls and steam passages would be chilled at the end of the stroke and, therefore, the live steam would be partially condensed, as it enters the cylinder, without doing work. It is in reducing this loss of steam by condensation, called initial condensation, that compounding effects economy in fuel, because if the expansion occurs successively in two cylinders, instead of all in one, the range of temperature is only one-half as great and consequently the condensation is reduced proportionately.

Q. What should be the relative sizes of cylinders in multiple expansion engines?

A. They should be so proportioned that approximately the same amount of work is done by each cylinder. The first cylinder will be the smallest in diameter and the last the largest.

Q. What names are given to the different cylinders of multiple expansion engines?

A. The one which takes the steam direct from the boiler is called the high-pressure cylinder, and the one in which it expands last before finally being exhausted to the atmosphere or condenser is called the low-pressure; the others are called intermediate-pressure cylinders.

Q. What is a receiver?

A. It is a chamber in which the steam is stored from the time it leaves one cylinder until it is admitted to the next.

Q. Why is a receiver necessary?

A. Because the cranks of the different cylinders are usually not placed in the same position. For example, in a two- or four-cylinder engine they would generally be placed at 90° and in a three-cylinder engine at 120°. Hence the cylinders are not taking steam during the time it is exhausted in the preceding cylinder and, therefore, a chamber must be provided for storing the steam until it can be used.

Q. Why are cranks set at different angles?

A. To secure a more uniform turning force on the crank shaft.

Q. Does not the fly-wheel accomplish the same result?

A. Yes; but if this can be done without the aid of a fly-wheel it is much better, especially since in many instances, such as in marine engines, a fly-wheel cannot be conveniently used.

Q. Why are compound engines operated as condensing engines wherever possible?

A. Because the increase in the mean effective pressure in the low-pressure cylinder is a large proportion of the total. Low-pressure cylinders of multiple expansion engines frequently have a mean forward pressure of only 3 or 4 pounds, and hence by the use of a condenser this may be increased very materially.

Q. What do you understand by a high-speed engine?

A. Strictly speaking, a high-speed engine is one

which has a high piston velocity; but the term is now generally used to mean engines of high rotative speed.

Q. What advantages do high (piston) speed engines possess as compared to low-speed engines?

A. Other things being equal, they are lower in first cost, more economical to operate, and run more smoothly.

Q. What additional advantage is possessed by high (rotative) speed engines?

A. They are better adapted for driving electric machinery and other shafting which requires to be run at a high speed of rotation.

Q. Why are high-speed engines lower in first cost?

A. The power of an engine depends on the piston area, stroke, mean pressure, and speed, varying directly as each one of these factors. If the speed is increased, any one of the other three factors may be proportionately decreased, and, therefore, it follows, that a high-speed engine may be built smaller and hence more cheaply for a given horse-power than a low-speed engine.

Q. Why are they more economical in the use of fuel?

A. Because one of the principal losses in steam engines is that due to initial condensation and re-evaporation, and this is the less the more steam passes through a given cylinder in a given time. Hence it is less in high- than in low-speed engines.

Q. Why do they run more smoothly?

A. Principally because the effect of the reciprocating parts is to equalize the turning force on the crank pin, so that it is nearly the same at every part of the stroke.

Q. What do you understand by automatic cut-off and throttling engines?

A. Automatic cut-off engines are those in which the speed is kept constant under a variable load by a governor acting upon the cut-off—that is, one in which the steam is admitted longer for heavy loads than for light loads, the exact point at which it is cut off being regulated by the governor. In the throttling engine, the period of admission remains the same under all loads, but the initial pressure is regulated by the action of the governor on a throttle valve.

Q. Which of the two is the more economical method?

A. The automatic cut-off; because when the pressure of steam is reduced by a throttle valve, it expands without doing work and hence an amount of energy is lost equal to that which would be necessary to raise the steam from the pressure at which it is admitted to the cylinder to that at which it is delivered by the boiler.

Q. Under what conditions could throttling engines be used?

A. When the load remains uniform or nearly so, because throttling engines with plain slide valves are simpler and cheaper to build than automatic cut-off engines.

Q. What are single- and double-acting engines?

A. Single-acting engines are those in which steam is admitted on one side of the piston only. In double-acting engines it is admitted alternately on either side of the piston.

Q. What are the relative advantages of these two types?

A. For the same diameter of cylinder, length of stroke, steam pressure, and speed, the double-acting engine develops twice as much power. The single-acting engine, however, has no piston rod, cross-head, or guides, the connecting rod being attached direct to the piston. Engines of this class usually run faster, however, than double-acting engines, and they are so arranged that the crank dips into a vessel filled with oil, every revolution, all of the moving parts being encased in an iron boxing. They are, therefore, well adapted for use where the atmosphere contains much grit and dust.

Q. What is a rotary engine?

A. It is one in which a motion of rotation is

produced directly by the pressure of the steam and not a reciprocating motion first, which is afterward converted into a rotary motion, as in the ordinary type.

VALVES AND VALVE GEARS.

Q. What do you understand by the valve gear of an engine?

A. All that part of its mechanism which is used in the distribution of steam.

Q. Of what does the simplest form of valve gear consist?

A. Of a plain slide valve, an eccentric, and the rods or links necessary for transmitting the motion of the latter to the former.

Q. Describe the plain slide valve.

A. The diagram on page 198 shows the simplest form of slide valve in its central position, that is, in the position where steam is neither admitted to nor exhausted from the engine. *V* is the valve, *S S* are the steam passages through which steam is admitted to the cylinder *C* from the steam-chest *X.* The latter, being in communication with the boiler, is always filled with live steam when the throttle valve is open. *E* is the exhaust passage which, being in communication with the exhaust pipe, allows the steam to pass into the atmosphere or condenser after it has done its work in the

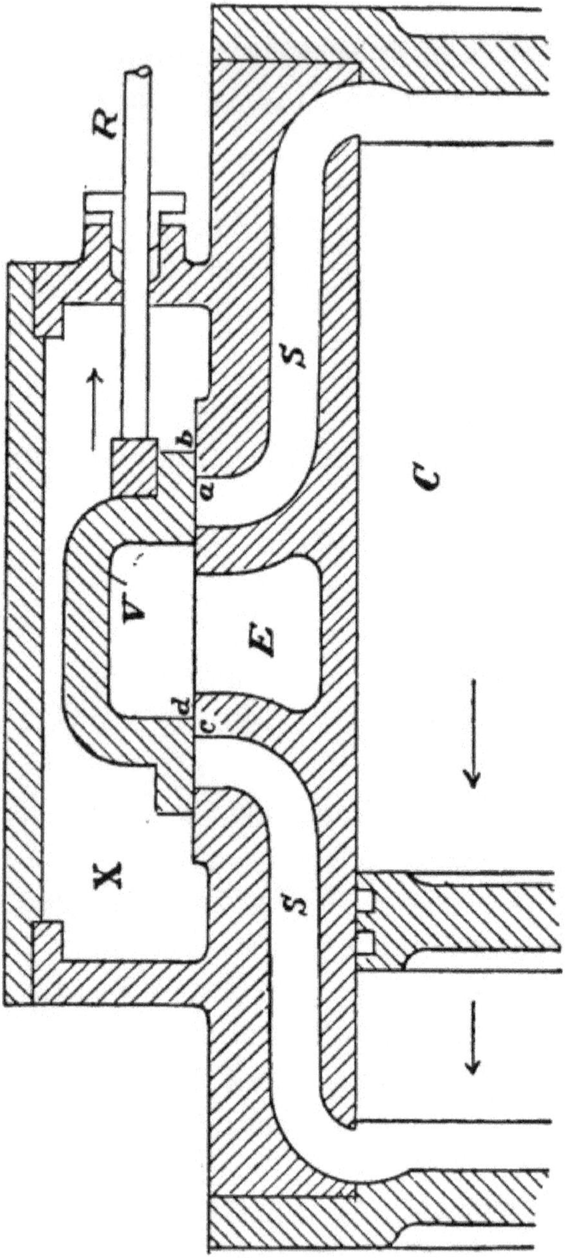

PLAIN SLIDE VALVE.

cylinder. *R* is the valve rod which receives its motion from the eccentric and, passing through a *stuffing-box*, imparts motion to the valve.

Q. Explain briefly the method of action of the valve.

A. As already stated, the valve in the above diagram is shown in a position where steam is neither admitted to nor exhausted from the cylinder. In this position of the valve, the piston which has nearly completed its stroke, is moving toward the left, while the valve is moving toward the right, as indicated by the arrows. Presently the valve will have uncovered the left steam passage and steam will be admitted behind the piston. This will continue until the steam passage is again covered by the valve on its return stroke. In the meantime the other steam passage will have been uncovered and placed in communication with the exhaust chamber *E*, and exhaust will take place until this passage is again covered by the valve. After that the process is reversed, steam being admitted to the right hand end of the cylinder and exhausted from the left; and so on, continuously.

Q. What are the four important events in the steam distribution, which take place in every double stroke of the engine?

A. Admission, cut-off, release, and compression.

Q. Explain what you mean by these terms.

A. When the passage is first uncovered *admission* takes place and continues until the point of *cut-off* is reached, which is when the passage is again covered. *Release* occurs when the passage is opened to the exhaust, and *compression* when the latter is closed. From the time steam is cut off until it is released *expansion* takes place.

Q. What do you mean by the terms lap, lead, eccentricity, travel, overtravel, angular advance?

A. *Outside* or *steam lap* is the distance the outer edge of the valve laps over the outer edge of the steam passage, in the central position of the valve, the distance *a b* in the cut.

Inside or *exhaust lap* is the distance the inner edge of the valve laps over the inner edge of the steam passage, in the central position of the valve, the distance *c d* in the cut.

Lead is the amount the steam port is open when the piston is beginning its stroke. If the piston begins its stroke before the steam passage is uncovered the lead is negative.

Eccentricity, or throw of the eccentric, is the distance from the center of the shaft to the center of the eccentric.

Travel of the valve is the total distance it moves on its seat between extreme positions. This travel is equal to twice the throw of the eccentric.

Overtravel is the distance the valve travels above what is necessary to fully uncover the steam passage.

Angle of advance is the angle by which the eccentric is in advance of the position which would bring the valve in its central position when the crank is on a dead center.

Q. Having given the various dimensions of a valve gear of this kind, how do you determine when the events described above will take place?

A. Graphically—that is, with the aid of some diagram such as Zeuner's, Sweet's, or Reuleaux's. Of these, Zeuner's is the one generally used in practice.

Q. Briefly explain the Zeuner diagram and its use.

A. *Draw a line *O X* to represent the crank at the beginning of the stroke, and with this as a radius draw the crank circle $X X_1$, X_2, X_3, X_4. Suppose the crank to turn in the direction of the arrow. Through the point *O* draw the line *R R′*, making the angle *R′ O Y′* equal to the angle of advance, and lay off the distances *O R* and *O R′* equal to the eccentricity or throw of the eccentric. On the lines *O R* and *O R′* as diameters draw the two circles *O C R D* and *O E R′ F*. With O as a center and a radius *O A* equal to the outside or

* From "Roper's Engineers' Handy-Book," pp. 391–393.

steam lap draw a circle *A C D*, and similarly with a radius *O B* equal to the inside or exhaust lap, draw a circle *B E F*. Through the point *O* and

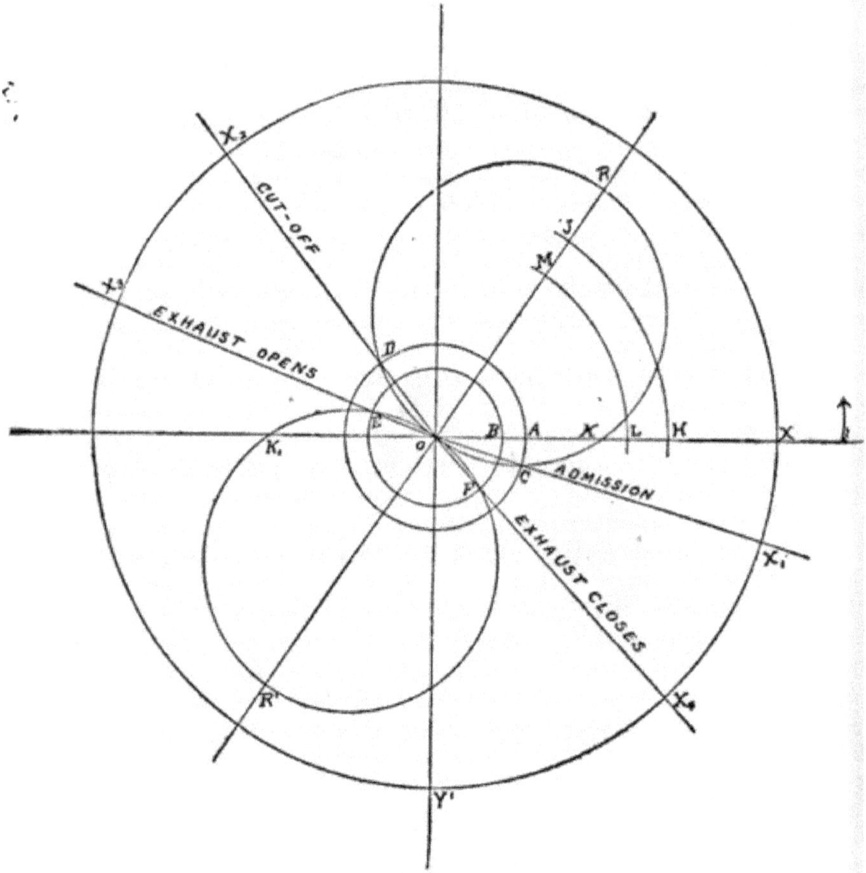

ZEUNER DIAGRAM.

the intersections *C*, *D*, *E*, and *F* draw the lines $O\,X_1$, $O\,X_2$, $O\,X_3$, and $O\,X_4$. We are now able to take from the diagram all of the data necessary for a complete understanding of the distribution of steam in the cylinder:

$O\,X_1$ is the position of the crank when admission of the steam begins.

$O\,X_2$ is the position of the crank when cut-off takes place, hence—

$X_1\,O\,X_2$ is the angle traversed by the crank during the period of admission.

$O\,X_3$ is the position of the crank when the exhaust opens.

$O\,X_4$ is the position of the crank when the exhaust closes, hence—

$X_3\,O\,X_4$ is the angle traversed by the crank during the period of exhaust, and—

$X_4\,O\,X_1$ is the angle traversed by the crank during the period of compression.

The distances from the intersection of the circles R and R' with the lines $O\,X$, $O\,X_1$, etc., represent the travel of the valve corresponding to the positions $O\,X$, $O\,X_1$, of the crank. The circle R represents the forward and the circle R' the return stroke, hence—

$O\,K$ is the distance the valve has traveled from its central position at the beginning of the stroke.

$O\,K'$, the same for the return stroke.

$O\,A$ is the outside or steam lap, hence—

$A\,K$ is the distance the steam port is open at the beginning of the stroke or the steam lead.

O R is the full travel of the valve.

O B is the inside or exhaust lap, hence—

B K is the distance the exhaust port is open at the beginning of the stroke or the exhaust lead.

At the points *C* and *D* the travel of the valve is just equal to the outside lap; hence in these positions of the crank the steam port opens and closes respectively; similarly at the points *E* and *F* the travel is just equal to the exhaust lap; hence, in these positions of the crank the exhaust port opens and closes respectively. If we lay down from the point *A* a distance *A H*, equal to the width of the port, and with *O* as a center and a radius *O H* draw an arc, cutting the line *O R* at *J*,—

J R is the distance the valve travels more than enough to fully open the port, or the over-travel.

Similarly, if we lay off from *B* the distance *B L*, equal to the width of the port, and from the center *O* and a radius equal to *O L* draw an arc, cutting the line *O R* at *M*,—

M R is the distance the valve travels more than enough to fully open the port to the exhaust.

It will thus be seen that by a careful study of the diagram all information necessary for the proper design and setting of the valve gear may readily be had. For example, in the above diagram the cut-off takes place a little later than $\frac{3}{4}$

stroke. It is evident that if it is desired to have the cut-off take place earlier, say at $\frac{1}{2}$ stroke, it will be necessary for the outside lap circle, *A C D*, to intersect the valve circle *R* in the line *Y Y'*. This may be accomplished by increasing the outside lap, by reducing the eccentricity, or by changing the angle of advance. However, any one of these changes would also affect the entire distribution, and it would probably be necessary to lay down several diagrams before the most advantageous dimensions could be obtained.

Q. How would you proceed to set the slide-valve of an engine?

A. Place the crank on the dead center and give the valve the necessary amount of lead; then turn the engine on the other center, and if the valve has the same amount of lead it is properly set. But if the lead on one end is more or less than on the other, the difference must be divided. When the valve is attached to the rod by means of jam-nuts great care must be taken not to jam the nuts against the valve, as that would prevent the valve from seating.

Q. What is a link motion?

A. It is a mechanism consisting of two eccentrics and rods and a slotted link, designed for the purpose of reversing an engine and varying its point of cut-off.

Q. How is this accomplished in the Stephenson link?

A. The two eccentrics, called respectively the forward and back eccentric, are placed on the shaft in different relative positions in such a way that, if the valve were operated by the one, the engine would move forward; and if by the other, it would be reversed. The link is attached to the ends of the two eccentric rods and hence receives a rocking motion. It is slotted and carries a movable block in the slot to which the valve rod is attached. If the block is at the end of the link nearest the forward eccentric, the engine will move forward, while if it is at the other end, it will be reversed.

Q. What happens when the block is in some intermediate position?

A. The travel of the valve becomes less as the block approaches the center, and hence the cut-off becomes earlier. In the central position of the block, the travel of the valve is not sufficient to uncover the ports, and hence the engine remains at rest.

Q. In the ordinary form of D slide valve, is there not a good deal of friction between the valve and its seat?

A. Yes; the friction in the old forms of slide valve is very great, because the steam pressure on the back of the valve forces it tightly against its seat.

Q. How can this be avoided to a great extent?

A. By the use of pressure plates, which relieve the back of the valve of its pressure, or by the use of the piston valve, which, being of circular cross-section instead of flat, is balanced and consequently the only pressure tending to force it against the seat is that due to its own weight.*

Q. What objection is there to piston valves?

A. It is claimed that the seat wears unevenly and hence they cannot be kept tight. With a suitable construction, however, the bushings forming the seat can be taken out and replaced with very little trouble and expense.

Q. Next to the slide-valve gear, as described above, what is the most common valve gear used in stationary engines?

A. The Corliss gear.

Q. What are the essential differences between the Corliss and the plain slide-valve gear?

A. Instead of a single valve which admits and exhausts the steam, the Corliss gear has four independent valves which rotate partially about an axis. The four valves, of which two are for the admission and cut-off and the other two for the release and compression of the steam in the cylinder, are operated by a single eccentric and wrist plate, but the two steam valves are connected

*See "Roper's Engineers' Handy-Book," pp. 398–402.

to the wrist plate in such a way that they can be detached at any moment. This is accomplished by a tripping or releasing mechanism controlled by a ball governor, and as soon as the steam valves are released, they are closed by the action of a dash pot, and hence the cut-off is under the direct control of the governor. The exhaust valves are not released from the wrist plate, and hence the release and compression are constant.

Q. What do you understand by a four-valve engine?

A. It is one having a valve gear midway between the plain slide valve and the Corliss gears. It has four independent valves like the Corliss, but, like the plain slide valve, their motion is *positive* and they have no releasing mechanism. The cut-off is varied by the travel of the valve.

Q. What are the relative advantages and disadvantages of the Corliss and four-valve types of valve gear?

A. The Corliss has the advantage that the cut-off is quick and sharp and that there is very little power lost in friction. The valves being, however, under the control of a spring or dash pot, they cannot be run at a high rotative speed. This constitutes the main advantage of the four-valve gear, that it can be run at as high a speed as a single-valve engine, and it is almost, but not quite, as economical as the Corliss. Both have the advantage over single-valve engines that the steam enters and leaves the cylinder by separate passages, and hence there is less loss by condensation. They are, therefore, much more economical in the use of steam than single-valve gears.

GOVERNORS.

Q. What are the principal methods in use for governing the speed of stationary engines?

A. By the centrifugal governor acting on the throttle valve—that is, by varying the initial pressure in the cylinder to suit the load; and by a centrifugal or inertia governor acting on the valve gear in such a way as to vary the point of cut-off to suit the load.

Q. Which is the better method, and why?

A. The one in which the cut-off is varied to suit the load, because it is much more economical in the use of steam, and the regulation is far better. Moreover, engines in which the steam pressure is throttled to suit the load often knock violently under light loads.

Q. Why should the steam never be throttled on engines running at a high piston velocity?

A. Because the force necessary to accelerate the reciprocating parts at the beginning of the stroke is so great in high-speed engines that if the steam

CENTRIFUGAL BALL GOVERNOR.

were throttled the fly-wheel would have to supply it, and hence there would be a reversal of pressure on the crank pin each stroke. This would not only cause very noisy running, but it would soon wear out the engine.

Q. How is the governor usually made to vary the cut-off?

A. By a releasing mechanism, as already explained above (Corliss valve gear); by the action of a ball governor on the block of a link, as in the Porter-Allen engine; or by a shaft governor.

Q. What is a shaft governor?

A. It is one in which the centrifugal action of a weight or weights, placed in a fly-wheel, is balanced against a spring or springs. The weights are attached to pivoted arms, and these in turn to the eccentric of the valve gear. As the speed increases, the tendency is for the weights to move away from the shaft and in so doing to alter the position of the eccentric, varying its angular advance or its throw, or both, and in this way altering the point of cut-off.

Q. What is the difference in the effect on the steam distribution when the cut-off is varied by the angular advance and by the throw of the eccentric?

A. If the angle of advance only is altered, the lead will increase as the cut-off is decreased. If the throw of the eccentric only is altered, the reverse takes place. Hence, in order to keep the lead constant with a single valve, both the throw

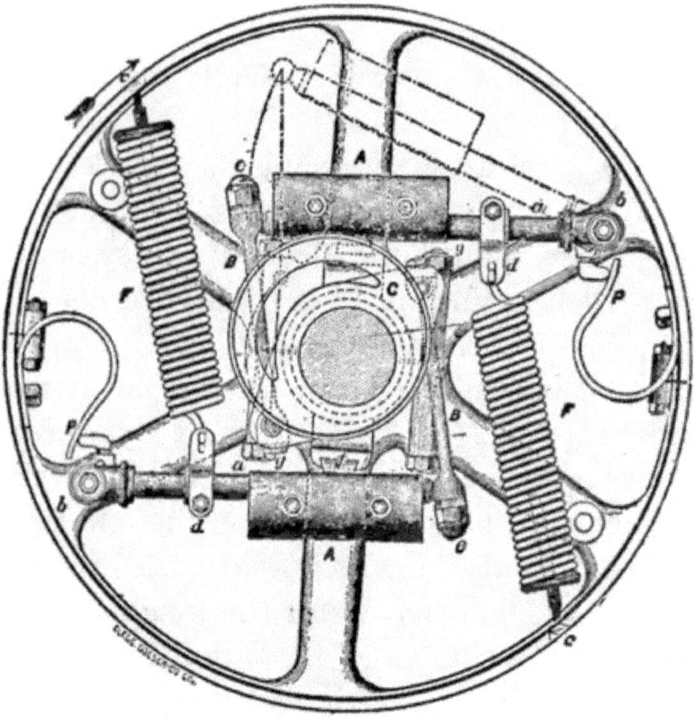

SHAFT GOVERNOR,—BUCKEYE TYPE.

(*A A* are the weights attached to the ends of arms *a a*. The arms are pivoted to the fly-wheel at one end and attached to the loose eccentric *C* at the other. *F F* are the springs which resist the tendency of the weights to move away from the shaft. In this type of governor the angular advance only is varied.)

of the eccentric and the angular advance should be varied. In the governor illustrated above, this is not necessary, because a separate valve is used to cut off the steam.

Q. How do you calculate the proper diameter for ball-governor pulleys?

A. To find the diameter of governor shaft-pulleys: Multiply number of revolutions of engine by diameter of engine shaft-pulley, and divide product by number of revolutions of governor.

To find diameter of engine shaft-pulley: Multiply number of revolutions of governor by diameter of governor shaft-pulley, and divide product by number of revolutions of engine.

INSTALLATION, CARE AND MANAGEMENT.

Q. What is the best material for engine foundations?

A. They should be of hard-burned brick laid in Portland cement or of concrete.

Q. How deep should they be carried?

A. The proper depth depends on the size of the engine. The builders usually furnish a foundation plan showing minimum depth, but they should always rest on solid ground.

Q. How should the foundation bolts and anchor plates be placed in the foundation?

A. A template should first be constructed to hold the bolts in their proper positions and the bolts suspended from the template. The bolts should be threaded at both ends and the lower nut held in a suitable pocket in the anchor plate. In building the foundation a space should be left around each bolt, sufficient to allow the bolt to be moved a half inch in any direction.

Q. How should the foundation be finished?

A. A cap-stone of granite makes the best finish, but, as a rule, the expense is too great. After the engine is set on the foundation and leveled by means of iron wedges, the space between the bottom of the engine and the top of the foundation should be filled with grout or, preferably, molten sulphur, to give an even bearing.

Q. Should foundations be built the same width from bottom and top?

A. No; they should be wider at the bottom and have a slope or batter of about two inches to every foot of height up to the floor-level. The top should be about an inch wider than the bed plate of the engine.

Q. How would you proceed to set up an engine?

A. First. Determine the position or location the engine is to occupy in the shop or factory.

Second. Lay out the line of the main shafting in the building, if there be any; if not, the line of the building itself, at, at least, three different points in the direction in which the main shafting is to run; now line down from the center of the main shaft, or from the line of the building, at two different points, to the floor on which the engine is to stand, and from these points line to the engine-shaft.

Third. Determine the height the bed-plate is to stand above the floor; also the depth of the foundation.

Fourth. Make a template the exact counterpart of the bed-plate, in which to hang the foundation bolts, and set this upon four props at right angles to the main shaft in the building.

Fifth. Lay up the brick foundation to the level at which the engine is intended to stand; then remove the template, and lower the bed-plate on to the foundation.

Sixth. Level the bed-plate by means of iron wedges and pour in sulphur to give it an even bearing. After that the nuts may be screwed down on the foundation bolts.

Seventh. A line should now be drawn exactly through the center of the cylinder, and another line through the center of the main bearing. This line will give the location of the pillow-block or outboard bearing.

Eighth. Place a straight edge across the bottom of the bearings and adjust them with the aid of a spirit level until they are perfectly level.

Ninth. Swing the fly-wheel into its proper position, slip the shaft through it and key it in place. Screw down the caps of the pillow-blocks.

Tenth. Place the cross-head, connecting rod, etc., in position, bolt on the front cylinder head, and adjust the valve gear.

Q. What are the principal points which should be kept in mind in running the steam and exhaust pipes for an engine?

A. They should be run in such a way that the free flow of steam will never be impeded. The steam- and exhaust-pipes should never be smaller than the outlets provided on the engine. The pipes should be run as straight as possible. Horizontal runs should be slightly inclined to allow the condensation to drain off in the same direction as the flow of the steam. The piping, if long, should have a suitable provision for expansion, and all steam- and exhaust-piping should be covered with some non-conducting pipe-covering.

Q. What is the first duty of an engineer in regard to the steam engine?

A. He should always keep it clean and free from rust, oil, and grit. This does not involve a great deal of labor, and adds very materially to the life of the engine.

Q. How should an engine be started?

A. First see that the drips are all open. The cylinder should then be warmed by slightly opening the throttle.

Q. How should the drips be left when the engine is not running?

A. They should be left open so as to allow the condensed steam to escape.

Q. How do you pack stuffing-boxes?

A. Before packing the piston- and valve-rods all the old packing should be carefully removed. The new packing should be cut in suitable lengths, and the joints placed at opposite sides of the box. The stuffing-box should then be screwed up until the leakage around the rod is stopped, and no further, as any unnecessary tightening of the stuffing-box will greatly diminish the power of the engine and soon destroy the packing by the increased friction. Piston-rod packing should always be kept in a clean place, as any dust or grit that may become attached to it has a tendency to cut or flute the rod.

Q. What precautions should be taken with the piston?

A. The spring packing in the cylinder should always be kept up to its proper place, because if allowed to become loose, the leakage materially reduces the power of the engine. Setting out packing-rings requires the exercise of great care, because, if set too tightly, the friction produced will not only have a tendency to cut the cylinder, but will also perceptibly lessen the power of the engine. The piston should be removed from the cylinder at least twice a year, and the joints formed by the rings on the flange of the head and the follower-plate carefully ground with emery and oil. If badly corroded, they should be faced up in a lathe and made perfectly steam-tight.

Q. How should the spindle of a ball governor be packed?

A. Great care should be taken, when packing the spindle of a governor, not to screw the packing down too tightly, as that would interfere with the free movement of the governor. All the parts of the governor should be kept perfectly clean and free from the gum formed by the use of inferior qualities of lubricating oils.

Q. How should the engine be lubricated?

A. All the surfaces subjected to friction should be provided with sight-feed oil-cups. These should be turned on as soon as the engine is started and examined at frequent intervals, to see that the supply is not exhausted and to make sure that every cup is feeding correctly.

Q. Is it advisable to use as much oil as possible on an engine?

A. No more oil should be used on an engine than is absolutely necessary, as it is not only a loss, but often detracts from the appearance of the engine, and greatly interferes with its free and easy movement, from the accumulation of gum and dirt on its working parts.

Q. Suppose any part of the engine should heat, what would be the proper thing to do?

A. First examine the lubricator, and if it is found that the heated part has not been receiving the proper amount of oil, the trouble can usually be remedied by giving it a liberal supply. Sometimes it is necessary in a new engine to keep the bearings cool, temporarily, with ice, although if they run very hot it is generally better to stop the engine if possible and determine the cause. In case the crank-pin should heat—which is a common occurrence with engines having a narrow bearing on the pin, but more particularly with engines that are slightly out of line—remove the key and slacken the strap and box; then pour in some flour of sulphur with a liberal supply of oil; then adjust the key, and the trouble will generally disappear. If the pillow-blocks of an engine should heat badly, remove the cap and pour in a good supply of pulverized bath-brick and water while the engine is in motion; after doing this for some time, wash out with oil, and wipe the bearing clean with waste. In case any of the bearings of an engine should heat through the accumulation of matter deposited from the oil used, or sand, grit, or whitewash being dropped into the bearings, use a strong solution of concentrated lye with oil when the engine is in motion.

Q. Where should the tools and materials used about an engine be kept?

A. They should be kept in a clean place. Never set steam-packing, cotton-waste, tops of oil-cups, or anything that is to be used around the cylinder, valves, piston-rod, or bearings of steam engines, on the floor, as they will invariably pick up sand or grit, which injure the rubbing and revolving surfaces with which they come in contact.

Q. How should gum-joints be made?

A. If they frequently need to be taken apart, the gum should be well coated with pulverized chalk or soapstone before being placed between the flanges. This prevents it from adhering to the metal and being destroyed when the joint is broken.

Q. What does a clicking noise in the cylinder indicate?

A. It frequently indicates the pressure of moisture, and it can generally be stopped by opening the drip-cocks.

Q. What are some of the principal causes of knocking in steam engines and the appropriate remedies?

A. Knocking in engines generally arises from the following causes:

First. Lost motion in the boxes on the cross-head, crank-pin, and the pillow-blocks, and in the key of the piston-rod in the cross-head. To stop it, take up lost motion by means of the key, or file off the edges of the boxes, if *brass-bound.*

Second. It is sometimes caused by the crank being ahead of the steam, which in most cases can be relieved by moving the eccentric forward in order to give more lead on the valve.

Third. Knocking is caused in many cases by too much lead on the valve. The simplest remedy for this is to move the eccentric back so as to give less lead.

Fourth. Frequently it is caused by the exhaust closing too soon. The best remedy for this would be to enlarge the exhaust-chamber in the valve.

Fifth. Insufficient clearance between the piston and the cylinder-head at the end of the stroke. The remedy for this kind of knocking would be to turn off the heads of the cylinder on the inside, so as to give more clearance.

Sixth. Knocking sometimes arises from the wrist of the cross-head and the crank-pin becoming worn out of *round.* The most effective remedy for this cause is to turn up the crank- and wrist-pin.

Seventh. Insufficient counter-bore in cylinder. In such cases the piston-rings wear a shoulder at each end of the cylinder, and whenever the keys are driven or the packing-rings set out, the edges strike these shoulders and cause the engine to knock. The most practical remedy for knocking arising from this cause is to *recounter-bore* the cylinder.

Eighth. Knocking is sometimes caused by the engine being out of line. The surest remedy for this kind of knocking would be to put the engine exactly in line.

Ninth. Sometimes it arises from shoulders becoming worn on the ends of the guides in cases where the gibs on the cross-head do not run over. The most reliable remedy for such knocking would be to replane the guides.

Tenth. Knocking is sometimes caused by the follower-plate being loose. The best preventive for such knocking is to bring the bolts up tight. To do so, it is sometimes necessary to remove the deposit of rust or grease in the bottom of the holes.

Eleventh. Very often it is caused by the packing around the piston-rod being too hard and tight. The most effectual remedy for that is to remove all the old packing from the box and replace it with new, and only screw the box up sufficiently to prevent the escape of steam. Too much friction on the rod is a great loss of power, and has a tendency to destroy the packing.

Twelfth. The knocking heard in the steam-chest is sometimes caused by lost motion in the jam-nuts or yoke that forms the attachment between the valve and rod. The remedy for this would be to remove the cover of the steam-chest and re-adjust the jam-nuts on the valve-rod.

ADJUNCTS OF THE STEAM ENGINE.

THE INDICATOR.

Q. What do you understand by the steam engine indicator?

A. An instrument which records the pressure in the steam cylinder at every point of the stroke.

Q. Give a brief description of the instrument and explain how this record is made.

A. The indicator consists essentially of a small hollow cylinder which communicates with the engine cylinder. A rod attached to the piston is enclosed in a spiral spring which presses against the piston and opposes its motion. The end of the rod extends through the cover at the top of the cylinder, and is attached to a series of levers, called a *parallel motion,* in such a way that a pencil attached to the end of the long lever will move in a vertical straight line when the piston ascends. A second hollow cylinder, carried on the same frame as the first, and called the paper drum, is mounted on a vertical spindle, about which it is free to rotate, but by the action of a spring contained in it the drum tends to remain in a fixed position. A groove, shown at the bottom of the drum, carries a cord which is attached by means of a reducing motion to some of the reciprocating parts of the engine, so that the pencil, when the engine is moving, would trace a horizontal line on the surface of the drum, which would represent the stroke of the engine. As the

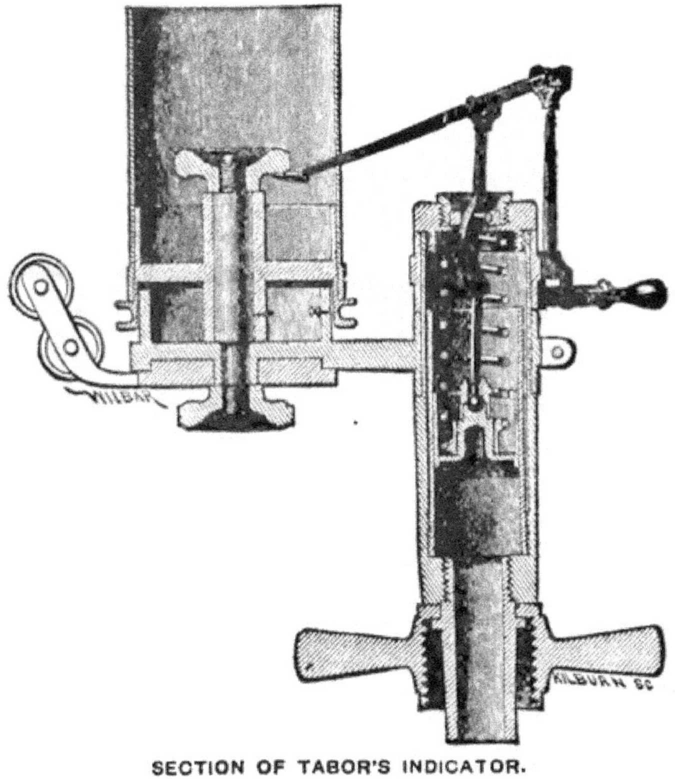

SECTION OF TABOR'S INDICATOR.

pencil, however, is moved up and down by the pressure of the steam in the cylinder, it follows that, if a paper is placed around the drum, a diagram will be traced, representing the pressure in the cylinder at every point in the stroke. The vertical height of any point in the diagram, from the bottom or *atmospheric line*, will represent the pressure, and the horizontal distances will represent the position of the piston.

Q. How would you proceed to take an indicator diagram?

A. It is impossible to give directions which would apply to all makes of indicators. I should carefully read the directions given by the makers of the particular type of instrument in my possession, and proceed accordingly.

Q. Sketch an indicator diagram and explain what it means.

A. In the accompanying diagram the line A A is the atmospheric line—that is, it is the line traced by the pencil on the paper when the engine is in motion before the indicator cylinder is placed in communication with the engine cylinder. Hence its position represents the pressure of the atmosphere. The point B represents the position of the pencil at the beginning of the stroke, and hence the vertical height B A of this point above the atmospheric line A A represents the initial steam pressure in the cylinder. The line B C represents the distance traveled by the piston during the period of admission, and the point C, where the first change in direction occurs, is the point of cut-off. Expansion now takes place in the cylinder and continues until the next change in direction occurs at D, which is the point at which the exhaust port begins to open. The steam is released from the cylinder, and the pressure falls more rapidly until the end of the stroke E, when it is about equal to that of the atmos-

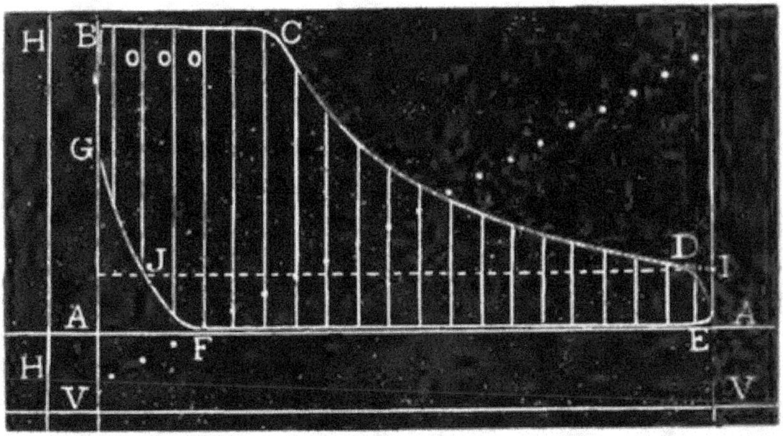

EXPLANATORY DIAGRAM.

phere. The piston then begins its return stroke against the back pressure represented by the vertical height of the line E F above the atmospheric line A A. If the engine exhausts into the atmosphere, this height is generally very small, while if it is a condensing engine, the back pressure line E F will be below the atmospheric line A A, indicating a negative back pressure. At F the exhaust closes and compression begins, which continues until the end of the stroke G. The same cycle is then repeated, and so long as the load, the initial pressure and the back pressure remain the same, the diagram traced by each successive stroke will be practically the same. For the other end of the cylinder the diagram will be similar but reversed.

Q. What are the principal things that may be ascertained about an engine with the aid of the indicator diagram?

A. The information furnished by the indicator diagram is of the most important kind. It enables us to determine:

First. The power of the steam engine under all conditions, or the power consumed by any one machine driven by the engine or by the engine itself in overcoming the friction of its parts.

Secondly. The forward and back pressure on the piston at any point in the stroke.

Thirdly. The average forward and back pressure and the mean effective pressure on the piston.

Fourthly. The positions of the piston when steam is admitted and cut off; the period of expansion, exhaust, and compression; the action of the valves; and, in fact, all questions relating to the steam distribution.

Q. How is the power developed by the engine, or the *indicated horse-power* calculated from the diagram?

A. The indicated horse-power of the engine is found by determining the mean effective pressure from the diagram and using it in the rules and formulæ for horse-power given on pages 177–180.

Q. Explain how to find the mean effective pressure.

A. There are two methods in common use,—one by the use of *ordinates* and the other by the *planimeter.* The latter method is more exact and less laborious than the former, but as a planimeter is not always available, the former method is much used, especially for rough calculations.

TO DETERMINE THE MEAN EFFECTIVE PRESSURE.

First Method.—Draw vertical lines A B and A I touching the ends of the diagram (see page 227), and apply a rule across them obliquely as shown by the dotted line in the diagram in such a way that some division on the rule, as $\frac{1}{16}$, $\frac{1}{12}$, $\frac{1}{10}$, or $\frac{1}{8}$, will divide the distance between the verticals just drawn an even number of times, preferably 20 times. Mark off points on this line, dividing it into equal parts excepting the first and last, which are only one-half as large as the intermediate spaces, and draw vertical lines or ordinates through these points, dividing the area enclosed by the diagram as shown. Next take a long strip of paper and apply its edge successively to each of the ordinates and mark their combined length on it. This length multiplied by the scale of the spring used and divided by the total number of ordinates will give the mean effective pressure. The length of the ordinates is measured between the forward- and back-pressure lines.

Second Method.—If a planimeter is used, it is only necessary to multiply the area enclosed by the diagram in square inches by the scale of the spring, and divide the product by the length of the diagram in inches. The quotient will be the mean effective pressure.

Q. What precautions must be taken if the indicated horse-power is to be calculated very accurately?

A. The mean effective pressure must be calculated separately from the diagrams of the head- and crank-ends of the cylinder. In doing this it must be remembered that the back-pressure line of one diagram belongs to the forward-pressure line of the other, and *vice versâ.* While in most engines in which the valves are properly adjusted the two back-pressure lines are identical, yet if the greatest accuracy is desired the mean effective pressure should be calculated by deducting from the mean forward pressure as obtained from the head-end diagram, the mean back pressure as obtained from the crank-end diagram, and *vice versâ.* It must further be borne in mind that the effective area of the piston at the crank end is less than that at the head end by the area of the piston rod. Hence the horse-power is different for the two ends and should be calculated independently; the total horse-power of the engine being equal to the sum of the two.

Q. Suppose it is desired to find the horse-power of an engine where the following dimensions and data are known:

Stroke = 36 inches,
Diameter of cylinder = 24 inches,
Speed = 150 revolutions per minute,
Diameter of piston rod = 4 inches.

The engine having been indicated with a spring whose scale was 60 pounds per square inch, it was found with the aid of a planimeter that the areas of the diagrams were as follows:

Head end = 3.54 square inches,
Crank end = 3.42 square inches,
Length of diagrams = 3.27 inches.

Calculate the mean effective pressures and the horse-power of the engine.

A. The mean effective pressure, according to the above (second) method, is—

$$\text{Head end, } \frac{3.54 \times 60}{3.27} = 64.95 \text{ pounds,}$$

$$\text{Crank end, } \frac{3.42 \times 60}{3.27} = 62.32 \text{ pounds.}$$

The area of the piston is—

$$.7854 \times 24 \times 24 = 452.39 \text{ square inches,}$$

and the area of the piston rod is—

$$.7854 \times 4 \times 4 = 12.57 \text{ square inches.}$$

Hence the effective areas of the piston are—

Head end,	452.39	square	inches.
	12.57	"	"
Crank end,	439.82	"	"

The total mean pressures on the piston are—

$$\text{Head end, } 452.39 \times 64.95 = 29385 \text{ pounds,}$$

$$\text{Crank end, } 439.82 \times 62.32 = 27409 \text{ pounds.}$$

The piston speed is—

$$\frac{36}{12} \times 150 \times 2 = 900 \text{ feet per minute;}$$

and therefore the horse-power—

$$\text{Head end. } \frac{29385 \times 900}{33000} = 801.4$$

$$\text{Crank end, } \frac{27409 \times 900}{33000} = 747.5$$

Total, 1548.9

CONDENSERS.

Q. What do you understand by a condenser?

A. An apparatus for condensing the exhaust steam of an engine, thereby reducing the back pressure and therefore increasing the power.

Q. How is this done?

A. By bringing the steam under the influence of cold water, either by bringing the two in direct contact or by allowing the steam to pass around a series of tubes through which the water flows. Condensers constructed on the first-named plan are called *jet condensers*, while the latter are termed *surface condensers*.

Q. What are the principal advantages and disadvantages of the two types?

A. Surface condensers have the advantage that the condensed steam is not mixed with the condensing water. Hence they are generally used on shipboard so that the condensed steam may again be used in the boilers. The vacuum is also generally higher in surface than in jet condensers, but they have the disadvantage of being heavier and much more expensive to construct than jet condensers. The tubes are also liable to become leaky and impair the vacuum.

Q. At what temperature should jet condensers be kept?

A. About 100° Fahr., at which temperature they have been found to operate most efficiently.

Q. What degree of vacuum should exist in a good condenser?

A. From 20 to 26 inches.

Q. What do you mean by 26 inches of vacuum?

A. As the atmospheric pressure will support a column of mercury about 30 inches in height, each inch of the mercury column would be equivalent to a pressure of about ½ pound. A complete vacuum (which can never exist) would be a vacuum of 30 inches, corresponding to a pressure of 0 pound per square inch; 20 inches of vacuum would be one-third less vacuum or one-third of the atmospheric pressure—that is, 5 pounds per square inch absolute pressure. Hence to find the absolute pressure in pounds per square inch, deduct one-half of the vacuum in inches from the pressure of the atmosphere. Thus 15 inches of vacuum would be, $15 - 15 \times \frac{1}{2} = 7\frac{1}{2}$ pounds per square inch absolutely.

Q. How much power is gained by the use of the condenser?

A. From 20 to 30 per cent., depending on the type and size of the engine.

Q. How much water is required for condensers?

A. About 25 times the quantity evaporated in the boiler.

TABLE

SHOWING VACUUM IN INCHES OF MERCURY AND POUNDS PRESSURE PER SQUARE INCH.

Mercury.	Pounds.	Mercury.	Pounds.
2.037	1	16.300	8
4.074	2	18.337	9
6.111	3	20.374	10
8.148	4	22.411	11
10.189	5	24.448	12
12.226	6	26.485	13
14.263	7	28.552	14

MATERIALS AND THEIR PROPERTIES.

Q. Of what is all matter made up?

A. Of chemical elements.

Q. What are chemical elements?

A. Substances having certain definite and peculiar properties which, so far, chemists have not been able to split up into simpler substances, and which it is presumed cannot be further split up.

Q. What are some of the elements?

A. Among the metals: Iron, Copper, Lead, Tin, Zinc, Silver, Gold, and Platinum. Among the non-metals are: Antimony, Bismuth, Silicon, Sulphur, and Carbon. Among those which exist normally in the gaseous condition are: Hydrogen, Oxygen, Nitrogen, and Chlorine.

Q. What are the substances called which are made up by the chemical combination of two or more elements?

A. Compounds, as, for example, Water, which is a compound of Oxygen and Hydrogen; Ammonia, which is a compound of Nitrogen and Hydrogen; Carbonic Acid, which is a compound of Carbon and Oxygen; Zinc Oxide, which is a compound of Zinc and Oxygen; and common Salt, which is a compound of Sodium and Chlorine.

Q. What are the *molecules* of a substance?

A. The smallest particles into which a substance can be divided without these particles losing any of the distinctive properties of the substance.

Q. Have you any idea as to whether molecules are visible under the microscope?

A. They are not. Were the magnifying power in any way much increased, they would still be too small to be seen. Our ideas as to their existence are derived not from sight, but from a variety of chemical phenomena.

Q. Is it conceived that there are particles even smaller than molecules?

A. Yes, the so-called atoms. It is believed that each molecule of a *compound* substance is made up of the atoms of the elements contained in the compound. For example, the molecule of salt is supposed to be made up of an atom of sodium joined to an atom of chlorine, and the water molecule is supposed to be made up of two hydrogen atoms joined to one oxygen atom. The molecules of the *elements* are supposed to be made up of two or more atoms of that element.

Q. What is meant by the term "atomic weight" of a substance?

A. It is found experimentally that the elements combine with each other in certain fixed proportions or in multiples of them. The figures which represent these proportions (hydrogen being used as the standard and its combining weight called "one") are called the *atomic weights.* For example: Experiment shows that hydrochloric acid is made up of 35.4 parts by weight of chlorine to 1 part by weight of hydrogen; and that in other chlorine compounds the proportion of chlorine is represented either by 35.4 or by some multiple of it, as 35.4×2, 35.4×3, etc. Thus, salt is made up of 35.4 parts by weight of chlorine to 23 parts by weight of sodium.

Q. What is supposed as to the construction of substances according to the molecular theory?

A. Every substance is supposed to be made up of an immense number of molecules, which, even in the solid state, are never entirely at rest, and in the gaseous state are in perpetual violent commotion, rushing about in straight lines in all directions with enormous rapidity.

Q. What are the principal properties of metals?

A. Their *malleability*, or capability to stand hammering; their *ductility*, or power of being drawn out into wire; their *tenacity*, or strength; their *hardness;* their *fusibility*, or ease of melting; and their relative weight, or *specific gravity.*

Q. Name some of the most malleable of the common metals.

A. Gold, Silver, Aluminum, Copper, Tin, Lead.

Q. Name the most ductile.

A. Platinum, Silver, Iron, Copper, Gold.

Q. What are some of the strongest?

A. Iron, Copper, Aluminum, Platinum, Silver.

Q. What are some of the least fusible?

A. Platinum, Iron, Copper.

Q. What are some of the heaviest, or which have the greatest specific gravity?

A. Platinum, Gold, Lead, Copper, Iron.

Q. How would you define the specific gravity of a substance?

A. The ratio of its weight to the weight of an equal bulk of water.

Q. How would you find the specific gravity of a solid body?

A. If it is heavier than water, weigh it in air and then weigh it suspended in water. The difference in weight is the weight of an equal bulk of water. Divide the weight in air by the weight of the equal bulk of water and the quotient is the specific gravity.

If the body floats put just the weight on it that is necessary to make it sink even with the surface of the water. Then from the sum of this weight and the weight in air subtract the weight in water. The difference is the weight of an equal bulk of water. Divide the weight in air by this and the quotient will be the specific gravity.

Q. How would you measure the specific gravity of a liquid?

A. Take a vessel filled with it and weigh it. Then weigh the same vessel filled with water. Divide the weight of the substance by the weight of the water and the quotient will be the specific gravity.

Q. Is there any simple instrument for testing the specific gravity of liquids?

A. Yes; the hydrometer, which consists of a graduated tube of small diameter attached to a bulb containing air enough to make it float. Just below this air chamber is a small bulb containing enough mercury to keep the apparatus upright. The graduations on the tube give the specific gravity of the liquid in which the hydrometer is placed.

Q. Is water used as the standard of specific gravity for gases?

A. No; air at a standard temperature of 32° Fahr. and at a pressure corresponding to the atmosphere at sea level.

COMMON METALS.

Q. What are the varieties of iron?

A. Wrought iron, cast iron, and malleable iron.

Q. What is steel?

A. A modification of iron, it being a combination of iron with varying percentages of carbon.

Q. What are some of the properties of wrought iron?

A. It is tough, malleable, ductile, fibrous, and can be welded.

Q. How does cast iron differ from wrought iron?

A. It contains carbon, sulphur, silícon, phosphorous and other impurities. It is crystalline in structure, is neither malleable, ductile, nor tenacious, but has the very important property of allowing itself to be cast.

Q. What is malleable iron?

A. Cast iron annealed amid iron oxides.

Q. What are its properties?

A. It is much more ductile than cast iron and has a higher tensile strength, though far inferiör in both respects to wrought iron and steel.

Q. What are the properties of steel?

A. Steel partakes of the properties of both wrought and cast iron, as some steels can be cast and others welded. By varying the percentage of carbon in its composition its characteristics can be widely changed. It can be made soft and ductile or hard and brittle. Steel also has the important property of *tempering*, or being artificially hardened by sudden changes of temperature.

Q. What effect on the strength of steel does an increase of the percentage of carbon have?

A. It increases the strength of steel.

Q. What effect does it have on the ductility of steel?

A. The ductility is diminished.

Q. At about what temperature is iron red hot?

A. At about 1000° Fahr.

Q. At about what temperature does iron melt?

A. At about 3000° Fahr.

Q. How much is iron expanded when its temperature is raised from freezing point to boiling point?

A. About $\frac{1}{900}$ of its length.

Q. What is the effect of a rise of temperature on the strength of iron?

A. It increases nearly $\frac{1}{7}$ to about 600° Fahr., after which it falls. At 1000° Fahr. its strength is about half the maximum.

Q. How does copper compare with iron in its principal qualities?

A. It is more malleable and more ductile. Its tensile strength is a little less than one-half. Its specific gravity is a little greater. It is a much better conductor for heat and electricity, its electrical conductivity being about six times that of iron.

Q. How is the tensile strength affected by heat?

A. It is diminished, disappearing entirely at about 1300° Fahr.

Q. What is the temperature at which copper melts?

A. At about 2000° Fahr.

Q. In what form is copper mostly used?

A. In the form of sheets and wires.

Q. In what other ways is it largely used?

A. In combination with other metals forming alloys.

Q. What are some of the principal alloys?

A. Brass, Bronze, and German Silver.

Q. What is the composition of brass?

A. It varies with the purpose for which it is to be used. Ordinary brass in foundries consists of 2 parts copper to 1 part zinc. A little tin or lead is sometimes added, but essentially brass is an alloy of copper and zinc.

Q. What is bronze?

A. Bronze is essentially an alloy of copper and tin, consisting of about 8 parts copper to 1 part tin.

Q. What is German Silver?

A. An alloy of copper and zinc, having a composition of about 3 parts copper to 1 part zinc.

Q. What are some of the striking properties of lead?

A. Its softness and malleability and its lack of elasticity. A very valuable property is that it is not readily oxidized nor attacked by acids.

Q. For what purposes is it largely used?

A. In sheets, pans, and pipes and as a constituent of paints.

Q. How does it compare, in tensile strength, with iron?

A. Its tensile strength is very small indeed in comparison with that of iron.

Q. What is its melting point?

A. About 600° Fahr.

Q. What is its specific gravity?

A. About 11, nearly double that of iron.

STRENGTH OF MATERIALS.

Q. What do you understand by the breaking strength of a substance?

A. The force, in pounds per square inch, that must be exerted to break a specimen of that substance when it is placed in a suitable testing machine. The breaking strength may be either tensile or compressive.

Q. What is the tensile strength?

A. The number of pounds necessary to pull asunder the test piece of 1 square inch cross-section, the force being applied in a line perpendicular to the plane of the section.

Q. What is the compressive strength?

A. The number of pounds that must be applied to crush the test piece.

Q. What is the tensile strength of cast iron?

A. About 16,000 pounds per square inch.

Q. What is the compressive or crushing strength?

A. About 100,000 pounds.

Q. What are the tensile and compressive strengths of wrought iron?

A. They are about the same, viz., 50,000 pounds.

Q. What can you say of the strength of steel?

A. It may be made to have almost any value, by varying the composition, from 50,000 to 200,000 pounds per square inch. The great increase in strength is accompanied by brittleness.

Q. What are the strengths of oak and pine?

A. Tensile about 7000 pounds and compressive about 3500 pounds per square inch.

Q. In calculating the sizes of pieces, either metal or wood, are the above figures used without any allowance for uncertainties?

A. No; we make use of what is termed a *Factor of Safety.* We assume that the load coming on the piece is a certain number of times greater than it really is and calculate the size of the piece accordingly. The ratio between the assumed load and the real load is the *Factor of Safety.*

Q. What values are used for the factor of safety?

A. This depends entirely upon the nature of

the load. If it is steady, with no vibration as in the roofs of houses, the factor is taken as *three.* When the load is fairly uniform, but with vibration, as in the case of shafting hung from the roof trusses, the factor should be *four.* If the direction of the load is reversed, putting the piece in alternate tension and compression, the factor should be *six.*

Q. Suppose it were desired to hang a weight of 50,000 pounds on the lower end of a wrought-iron rod. What should be the area of the cross-section of the rod?

A. This is a case of a steady load where the factor of safety to be used is *three.* Multiplying the actual load by 3 we obtain 150,000 pounds as the load to be assumed. The tensile strength of wrought iron being about 50,000 pounds per square inch, it is evident that we must have a section of 150,000 ÷ 50,000, or 3 square inches.

Q. On what does the weight that a beam will support, depend?

A. On the length of the beam between the points of support, on its width and depth, and on the manner of application of the load.

Q. What difference does it make as to the manner of loading the beam?

A. It will support a much greater load if it is uniformly loaded than if the load is applied at one point.

Q. What do you mean by a uniformly loaded beam?

A. A beam is uniformly loaded when the weight per square inch resting on it is the same at all parts of its length.

Q. When a beam is supported at both ends, at what point will a given load break the beam most readily?

A. At the middle of the beam.

Q. What is the difference between the load which if applied in the middle will break a beam, and the load needed to break it if it is uniformly distributed?

A. A given beam will support a uniformly distributed load twice as great as that which will break it if it is applied at the middle.

Q. Can the values for crushing strength be safely used in all cases?

A. Not when the length of the piece in compression has a length greater than four times a diameter. When this is the case the piece becomes a column, and a bending action comes into play, causing the piece to break long before the load corresponding to the compressive strength has been reached.

ELECTRICITY.

Seven simple experiments contain the fundamental principles on which nearly all electrical apparatus depends.

Experiment 1.—Place in a jar containing a solution of chromic acid a plate of zinc and a plate of carbon. The plates should be near each other without actually touching, and each should have fastened securely to it a short piece of small copper wire. Place in another glass jar a solution of copper sulphate and let the ends of the copper wires dip into the copper sulphate solution *without touching each other.*

Q. What will happen to that part of the copper wires dipping into the solution?

A. The wire attached to the carbon plate will be gradually eaten away, while the wire attached to the zinc plate will increase in size by an equal amount.

Q. What is deposited on this wire to increase its size?

A. Pure copper.

Q. Suppose this wire were made of some other material than copper, would copper be deposited on it?

A. Yes; if made of iron, zinc, lead, or carbon.

Q. What does this experiment seem to show?

A. That there has been set up a *current* of something which apparently carries copper along with it.

Q. What name has been given to this current?

A. The electric current.

Q. Could other plates than zinc and carbon be used to produce it?

A. Yes; though zinc is generally used for one of the plates.

Q. Could another solution than chromic acid be used?

A. Yes; the solution must be one which readily attacks one of the plates, and it is usually some strong acid.

Q. What is the apparatus called in which an electric current is produced by chemical action?

A. A battery cell, or, simply, a cell.

Q. What is a battery?

A. Properly speaking, a battery means several cells, but it is often used to mean simply one cell.

Q. What is the wire called to which copper is carried?

A. The kathode.

Q. What is the wire called from which copper is taken?

A. The anode.

Q. In which direction does the current flow in the copper sulphate solution?

A. From the anode to the kathode.

Q. Is there a current flow through the cell containing chromic acid?

A. Yes; resulting in taking zinc from the zinc plate and carrying it into solution.

Q. Suppose one of the copper wires were cut, what effect would this have on the flow of current?

A. It would stop completely the action described above.

Q. What does this show?

A. That what is called the electric current was flowing around through a path or *circuit*, starting, say, at the carbon plate, thence through the copper wire attached to that plate to and through the solution of copper sulphate, then through the other wire to the zinc plate, and finally through the chromic acid solution back to the carbon plate. Any interruption of this *circuit* stops the flow of current.

Q. Would pulling one of the wires out of the copper sulphate solution have the same effect as cutting the wire?

A. Yes.

Q. Of what electrical industry is this experiment the basis?

A. Electro-plating.

Experiment 2.—Pull the copper wires out of the copper sulphate solution and touch them together.

Q. What will be observed?

A. The wires become heated.

Q. Equally all along their length?

A. Apparently so.

Q. Is the zinc plate being dissolved as in Experiment 1?

A. Yes.

Q. What does this experiment show?

A. That the electric current heats bodies through which it passes.

Q. Suppose the wire connecting the zinc and carbon plates is made longer, what will occur?

A. The heating will be less.

Q. And if the wire is made shorter?

A. The heating effect is much greater.

Q. What would you infer from this?

A. Since a decrease in the heating means a decrease in the current, and since this was caused by lengthening the wire, it would seem that the wire opposes a resistance to the flow of the electric current, and that the longer the wire the greater the resistance which it offers.

Q. Can you think of any electrical apparatus working on the principle shown in this experiment?

A. Electric heaters and certain electric measuring-instruments.

Experiment 3.—Bring a compass needle or a freely suspended bar magnet near the wire in Experiment 2.

Q. What will be observed?

A. The magnet is evidently acted upon by some force due to the current flowing through the wire. After oscillating it comes to rest, pointing cross-ways to the wire and nearly perpendicular to it.

Q. Is this the case all along the wire?

A. Yes.

Q. Why does the needle not stand exactly perpendicular to the wire?

A. Because normally it tends to point north. The current through the wire tends to make it stand perpendicular to the wire. It actually takes a direction between these two.

Q. Notice which way the north-seeking pole of the magnet points. Now, if the magnet is held first above the wire and then below, what occurs?

A. Although the needle tends to stand in a direction cross-ways to the length of the wire, yet when above the wire the north-seeking pole points in one direction, and when below the wire in the opposite direction.

Q. Is there any rule for telling in what direction it will point?

A. Yes, one known as Ampère's rule, which is: "*Imagine yourself swimming* WITH *the current and turned either on your side, face, or back, so as to look at the magnet. Then the north-seeking pole of the magnet will point toward your left."*

Q. In the above experiment, suppose that the wire carrying the current is free to move while the magnet is fixed, what will occur?

A. The wire will move either toward or away from the magnet, according as one pole or the other of the magnet is presented to it.

Q. What does this show?

A. That there is a force existing between a magnet and a wire carrying a current, similar to the force existing between two magnets. Further experiment shows that the *strength* of this force depends on the nearness of the magnet to the wire carrying current, and that the *direction* of the force depends on the position of the wire with respect to the two poles of the magnet.

Q. Can this magnetic force be represented conveniently by lines as in the case of other forces?

A. Yes. We conceive that around every magnet or wire carrying current lines could be drawn either straight or curved, which at any point of their length should represent the direction of the resultant magnetic force at that point.

Q. How could you actually lay out the lines of force due to any magnet, say a bar magnet?

A. If we could obtain a north-seeking pole of a

magnet without its accompanying south-seeking pole, we could place it near the north-seeking pole of the bar magnet and observe the path which it pursued from the north pole to the south pole and plot this path on paper. We would then place the test pole at another point of the north pole of the bar magnet, and again observe the path and plot it, and so on. In this way the space around the magnet could be mapped out.

Q. What is the space around a magnet, in which magnetic force exists, called?

A. The *field* of that magnet.

Q. Does every magnet have a field?

A. Yes; and since lines of force could be drawn in this field which would represent the direction of magnetic force, we say that every magnet produces lines of force.

Q. What, then, is a line of force?

A. It is a line which represents the direction of magnetic force in the region where the line is drawn or may be supposed to be drawn.

Q. What is the positive direction of the line of force?

A. That direction in which a free north-seeking magnetic pole would move. A free south pole would move in the opposite direction.

Q. Since we cannot obtain a free north pole for testing the direction of magnetic force, how can we explore and map out the magnetic field due to any magnet or wire carrying current?

A. By taking advantage of the fact that a short magnet will, if free to move, place itself lengthwise along the lines of force.

Q. Explain how the experiment is performed.

A. Place under a piece of window-glass a bar magnet, and dust on the upper side of the glass some iron filings. These filings become magnets which are exceedingly short, and when they are jarred by tapping the glass they are free to move and set themselves into lines corresponding to the lines of magnetic force as shown in the cut.

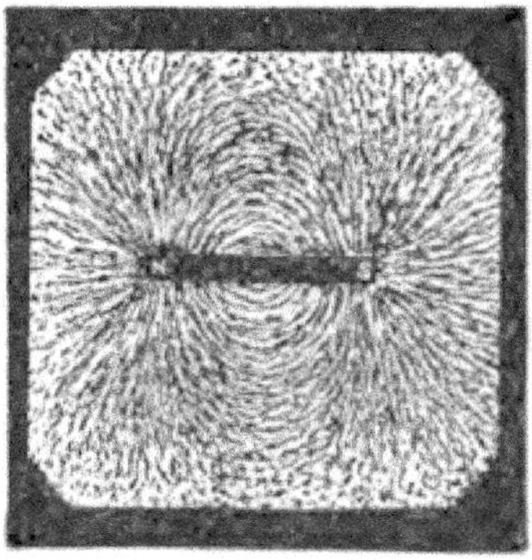

Q. Why are the lines of filings more dense at some points of the field than at others?

A. Because the strength of the magnetic force is greater at those portions of the field.

Q. How would you describe the lines of force due to a bar magnet?

A. As curved lines running from the north pole to the south pole.

Q. What are the lines of force due to a horseshoe magnet?

A. Principally straight lines from the north to the south pole.

Q. How can you obtain the field due to a current in a straight wire?

A. By drilling a hole in the piece of glass and passing the wire vertically through this hole and then dusting on iron filings.

Q. What are the lines of force due to a current in a wire?

A. Circles concentric with the axis of the wire, the positive direction being in the direction in which the hands of a watch move.

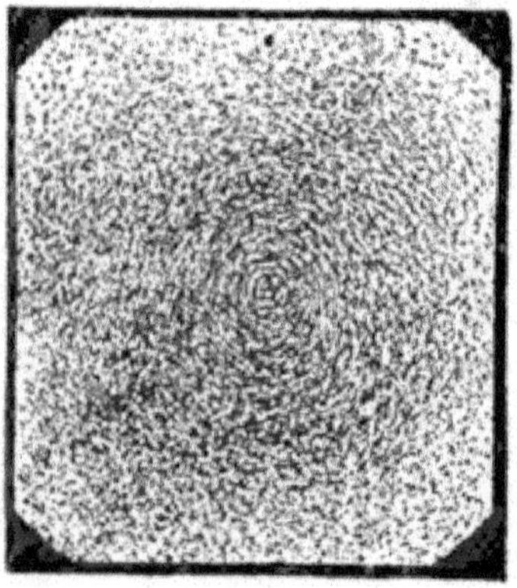

Q. Where is the magnetic force greatest?

A. Next to the wire, as shown by the greater density of the lines of force.

Q. Suppose the current through the wire were greatly increased, how would the density of the lines be affected?

A. It would be increased in the same proportion as the magnetic effect of the current is strictly proportional to the strength of the current.

Q. When a coil of wire carrying a current is brought near a magnet, can the direction of motion of the coil or magnet be told in advance?

A. Yes; they will move in such a way that the greatest possible number of lines of force due to the magnet will pass through the coil.

Q. For what practical purpose can this principle of the effect of an electric current on a magnet be used?

A. We can detect currents in wires by bringing a magnet near the wires, and can also, by applying Ampère's rule, determine in which direction the current flows.

Q. Is there any other method of determining the direction of flow of a current.

A. Yes; by making use of the principle illustrated in Experiment 1. The current can be led into a solution of copper sulphate (or nearly any solution of a metallic salt), and by noting which of the wires increases in size we can tell in which direction the current flows, as it flows *toward* the wire which has copper deposited on it.

Q. Can we increase the effect of the current on the magnet?

A. Yes, in three ways: By increasing the strength of current, by bringing the wire and the magnet nearer together, and by winding the wire which carries the current in a coil and placing the magnet in the axis of the coil.

Q. When this is done, what direction will the current in the coil tend to make the magnet assume?

A. A direction parallel to the axis of the coil. Since the magnet is also acted on by the earth's magnetism tending to make it point north, it will actually assume a position between these two directions. The angle which it makes with north depends on the relative strength of the earth's magnetic force and the magnetic force due to the coil. With no current passing through the coil the magnet points due north. When a small current passes through the coil the magnet is slightly deflected. A larger current deflects it more, and so on.

Q. What is the apparatus called which consists of the coil of wire and pivoted magnet described above?

A. A galvanometer.

Q. For what purposes should you say that the galvanometer would be useful?

A. For detecting the presence of electric currents, determining in which direction they flow and also to measure their strength.

Experiment 4.—Connect to a galvanometer, as described above, the terminals of an auxiliary coil of wire placed a few feet distant, the connection being made by leading a wire from one end of the auxiliary coil to one end of the galvanometer coil, and another wire from the other end of the auxiliary coil to the other end of the galvanometer coil. Bring a strong magnet near the auxiliary coil, watching at the same time the magnet needle of the galvanometer.

Q. What occurs?

A. The magnet needle gives a sudden jump and continues to oscillate to and fro, coming to rest a little while after the motion of the strong magnet has stopped.

Q. What does this show?

A. The jump of the galvanometer needle shows that an electric current has been produced by moving the magnet near the auxiliary coil. The fact that after the magnet stops the needle comes to rest in its original position, shows that the current is produced only while the magnet is moving.

Q. Suppose that instead of moving the magnet toward the auxiliary coil, the coil is moved toward the magnet?

A. The galvanometer needle jumps in the same direction as before, showing that current is produced in the same way and in the same direction.

Q. Suppose that the magnet and coil are moved away from each other?

A. The needle jumps as before, but in the opposite direction.

Q. What do you conclude from all this?

A. That moving a wire and a magnet relatively to each other produces an electric current, and that the direction of the current depends on the direction of the motion.

Q. Has the current so produced the same properties as the current produced by a battery?

A. Absolutely the same; the two are identical.

Q. What piece of electric apparatus is based on the principles illustrated by this experiment?

A. The dynamo.

Q. Making use of the idea of lines of force in the above experiment, what result do you arrive at?

A. Moving the magnet nearer the coil causes the coil to cut across lines of force due to the magnet, and since a current is produced by the motion we may conclude that *whenever an electric conductor cuts across lines of force an electric current is produced.*

Q. When the magnet was moved away there was a current produced in the opposite direction by the cutting of lines of force. Is there any convenient rule for determining the direction of the induced current?

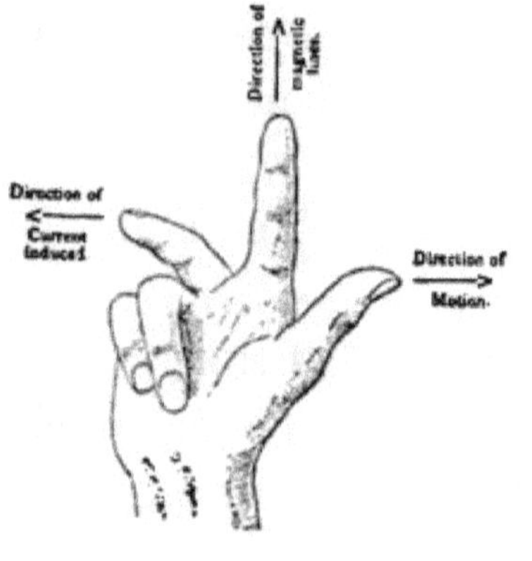

A. Yes; a rule known as Fleming's.

Point the forefinger along the positive direction of the magnetic lines and point the thumb stretched at right angles in the direction in which the conductor moves. If now the second finger be stretched at right angles to both thumb and forefinger, it will point in the direction of the induced current.

Q. When the magnet is moved nearer the coil, the number of lines of force due to the magnet, which is enclosed by, or which passes through, the coil, is increased, might we not say that a current is produced whenever the number of lines enclosed by a coil is changed?

A. Yes; and when the conductor is in the form of a coil this idea is of great value. Looking along the positive direction of the lines of force, when the number enclosed by the coil is increased, the current around the coil is left-handed as we look at it. If the number enclosed by the coil is diminished, the current will be right-handed as we look at it.

Q. What do you mean by right-handed?

A. In the direction in which the hands of a watch move.

Experiment 5.—If the current from a battery or other current generator be led through a wire which is coiled around a rod of iron, the iron becomes strongly *magnetized*, as we say; that is, it exhibits all the properties of a magnet. It attracts other pieces of iron, and it has polarity, one end attracting the north-seeking pole of a bar magnet and the other end repelling it.

Q. What is the combination of a piece of iron with a coil of wire around it called?

A. An electro-magnet.

Q. After current is cut off from the coil, does the iron still exhibit magnetic qualities?

A. Only feebly. The magnetism still remaining is called permanent or residual magnetism.

Q. What is the advantage of an electro-magnet over a permanent magnet?

A. For the same size the electro-magnet is much more powerful.

Experiment 6.—Suspend a coil of wire so that it can turn freely and lead a current through the wire. Then bring a magnet near it.

Q. Will the coil be affected by the magnet?

A. Yes, the coil will turn so as to enclose as many as possible of the lines of force due to the magnet and will finally come to rest in that position.

Q. Suppose the other pole of the magnet be presented toward the coil?

A. The coil will turn in the opposite direction and come to rest in such a position that it encloses the greatest possible number of lines of force due to the magnet.

Q. Suppose just at the moment the coil gets into the position of enclosing the maximum number of lines the current is reversed in direction, what will be the effect?

A. The coil will continue to turn in the same direction and will make a half turn, after which it will stop.

Q. Can you determine in which direction the coil will turn?

A. Yes, by applying Fleming's rule previously mentioned, using the left hand. Point the forefinger along the positive direction of the lines of force due to the magnet at any part of the coil. Point the second finger, held at right angles to the forefinger, in the direction of the current in that part of the coil. Finally, extend the thumb at right angles to both of the fingers. The direction in which the thumb points will be the direction in which that part of the coil will move.

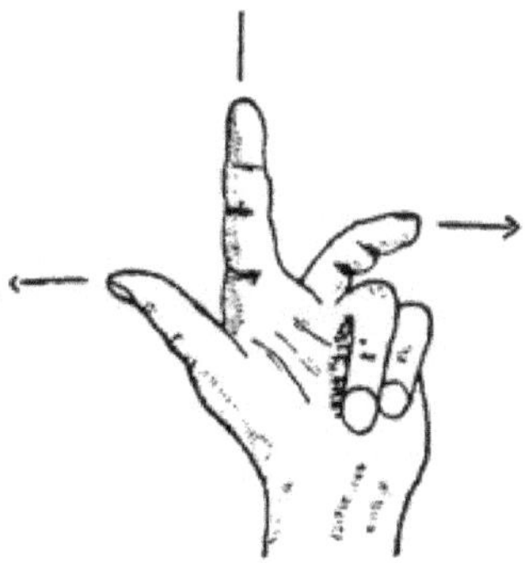

Q. And if at this point the direction of current is again reversed?

A. The coil will rotate in the same direction one half-turn further.

Q. What piece of well-known electrical apparatus operates in this manner?

A. The electric motor.

Q. Does it make any difference whether the magnet is a permanent or electro-magnet?

A. None at all, except that greater strength can be secured by using an electro-magnet.

Experiment 7.—Suppose we have the same coil of wire as in Experiment 6, which we will call coil No. 1, connected to a galvanometer, and near it a second coil attached to a battery. A current is flowing through coil No. 2, but not through coil No. 1, of course.

Q. What occurs if we suddenly disconnect the battery from coil No. 2, and what does it show?

A. The needle of the galvanometer will give a sudden jump, showing that by stopping the current through coil No. 2 a current has been produced, or induced, as we say, in coil No. 1, although coil No. 1 is not connected to coil No. 2 in any way. In a moment or two the needle of the galvanometer will come to rest at its original position, showing that the current has ceased.

Q. What will occur if the battery be again connected to coil No. 2?

A. The needle will again jump, but this time in the opposite direction, showing that the induced current is in the opposite direction.

Q. Suppose that the current instead of being entirely stopped were diminished and then increased, what would happen?

A. We should see the needle go first one way and then the other, as before, showing that any change in the strength of current in coil No. 2 tends to induce a current in No. 1.

Q. Looked at from the standpoint of lines of force, what has occurred in this experiment?

A. From the standpoint of lines of force, when the current in coil No. 2 is increased more lines of magnetic force are enclosed by No. 1, and a current is produced. When the current is diminished less lines pass through No. 1, and a current is induced in the opposite direction. The nearer the two coils are to each other the greater the effect, and if a soft iron core be introduced into the axis of the coils, the induced current becomes enormously greater than before.

Q. What electrical apparatus is illustrated by this experiment?

A. The transformer.

Experiment 8.—Connect a battery to a galvanometer and notice the reading of the needle which shows what current is flowing through the circuit. Connect in tandem another cell of battery.

Q. What will occur?

A. The reading of the galvanometer needle will be increased, being about double what it was before.

Q. What does this show?

A. That the current through the circuit is double.

Q. Has the resistance of the circuit been appreciably changed?

A. No.

Q. What could have caused double flow through the same resistance?

A. Reasoning from analogy to the flow of water, the pressure tending to cause flow must have been doubled.

Q. Would you then conclude that there is such a thing as electrical pressure?

A. Yes, and that each generator, as, for instance, a battery, furnishes a definite pressure, and that when two are connected in tandem the two together furnish a pressure which is the sum of the pressures furnished by each.

Q. What other names are there for electric pressure?

A. Difference of potential (P. D.), electromotive force (e. m. f.), and voltage.

Q. The battery produces electric pressure by means of chemical action; is there any other method?

A. Yes; an electric pressure is produced wherever a conductor cuts across lines of force; or if the conductor is in a coil a pressure is produced whenever the number of lines of magnetic force enclosed by the coil is in any way changed. The pressure continues only so long as the cutting or change of number of lines of force continues.

Q. Upon what does the amount of electric pressure depend?

A. On the rate of cutting the lines of force—that is, the number cut per second or the change per second in the number enclosed by a coil.

Q. Suppose a coil has 10,000 lines of force passing through it, its plane being perpendicular to the lines of force, which lines are in this case supposed to be parallel and straight. Now let the coil be rotated one quarter-turn, how many lines will it enclose?

A. Zero.

Q. Suppose it took one-quarter of a second to make the quarter-turn, what would be the rate of change of lines of force enclosed by the coil?

A. $10,000 \div \frac{1}{4} = 40,000$ per second.

ELECTRICAL UNITS.

Q. What is the unit of electrical pressure or electro-motive force?

A. The *volt,* which is the pressure furnished by a certain standard cell.

Q. What is the unit of resistance?

A. The resistance of a column of mercury 41.85 inches long and weighing 223 grains at 32° Fahr. It is called the *ohm.*

Q. Are the standard ohms and multiples of the ohm used in practice made of mercury?

A. No; they are made of German-silver wire, or an alloy of copper, nickel, and one or more metals.

Q. What is the unit of current?

A. It is the current which will deposit, in one second, on the kathode plate, from a standard solution of silver nitrate, .001118 gram (.017 grain) of silver. It is called the *ampère,* and is in its nature a unit of rate of flow and analogous to a flow of a certain quantity per second.

Q. What other common unit is employed?

A. The *watt,* which is the unit of power. It is equal to a volt-ampère; that is, the power in watts is equal to the product of the number of ampères flowing multiplied by the number of volts pressure causing the flow.

Q. What relation does the watt bear to a horse-power?

A. One horse-power equals 746 watts exactly, or, in round numbers, 750.

Q. What multiple of the watt is found convenient?

A. The *kilowatt,* written K.W., which is 1000 watts and nearly equal to $\frac{4}{3}$ horse-power.

Q. In measuring electrical properties, such as current, pressure, resistance, or power, what is the general method of going about the work?

A. Take current as an example. We find some effect of current easy to observe, and we agree to call a current which produces this effect to a certain extent unit current, as, for example, the current which in one second will deposit from a nitrate of silver solution .017 grain of silver is called unit current. Having an unknown current which it is desired to measure, we observe how many grains of silver it will deposit in one second, and if it deposits .17 grain we call it a current of 10 units or 10 ampères. Of course, no one in actually measuring a current now goes through the long process of measurement by means of depositing a metal any more than in order to measure a length he makes a journey to the British Museum to get the standard yard-stick. Convenient instruments working on the principle of a galvanometer are made so that when a current of 1 ampère flows through their coils their needle points to 1; with a current of 2 ampères, points to 2, and so on.

Q. What multiples of the units given above are in common use?

A. The megohm = 1 million ohms.
The microhm = 1 millionth part of 1 ohm.
The kilowatt = 1 thousand watts.

Q. Can these prefixes, *meg*, *micro*, and *kilo*, be used with the other electrical units?

A. Yes; although such use is not very common.

RESISTANCE.

Q. How is the resistance of a conductor affected by increasing its length?

A. The resistance is increased proportionately to the increase in length.

Q. What is the effect of increasing the area of cross-section?

A. The resistance is *lessened* proportionately; in other words, the resistance is inversely proportional to the area of the cross-section.

Q. A certain size wire, 100 feet long, has a resistance of 2 ohms,—what will be the resistance of 200 feet of the same wire?

A. 2 × 2, or 4 ohms.

Q. Suppose that 100 feet of wire $\frac{1}{10}$ inch diameter has a resistance of 1 ohm,—what would be its resistance if the diameter were $\frac{1}{20}$ inch?

A. Since the new diameter is one-half the old, the area of cross-section of the new wire is $\frac{1}{2} \times \frac{1}{2}$, or one-quarter that of the old wire. The resistance therefore would be four times *greater*, or 4 ohms.

Q. What is meant by the conductivity of a wire or other conductor?

A. The opposite of resistance. It is numerically equal to 1 divided by the resistance.

Q. A wire has a resistance of 100 ohms,—what is its conductivity?

A. $\frac{1}{100}$, or .01.

Q. When two resistances, as *Y* and *R*, are joined as shown in the figure, how are they said to be connected?

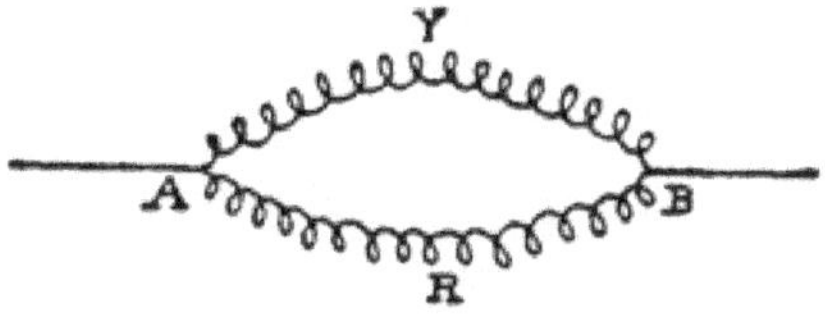

A. In parallel or multiple.

Q. When so connected, what is their joint resistance, that is, the resistance from *A* to *B*?

A. It is found by the formula, joint resistance $= \frac{R \times Y}{R + Y}$.*

Q. Two resistances of 10 and 20 ohms respectively are joined in multiple,—what is their joint resistance?

A. $\frac{10 \times 20}{10 + 20} = \frac{200}{30} = 6\frac{2}{3}$ ohms.

Q. When the resistances are equal, what is the joint resistance?

A. One-half the resistance of one.

Q. When several equal resistances are connected in multiple, what is their joint resistance equal to?

A. To the resistance of one divided by the number of them.

Q. When are two conductors said to be connected in *series?*

*For complete explanation, see "Roper's Engineers' Handy-Book," page 665.

A. When they are joined tandem, or end on.

Q. When two resistances are connected in series, what is their joint resistance equal to?

A. To the sum of the separate resistances.

Q. What is specific resistance?

A. It has the same relation to resistance that specific gravity has to weight. It is the resistance of a cubic inch, or it may be expressed in cubic centimeters.

Q. What are some of the substances having large specific resistance?

A. Of the metals—lead, mercury, and alloys. The non-metals have a much higher specific resistance.

Q. What are some substances having a low specific resistance?

A. Copper, silver, and gold.

Q. What are non-conductors?

A. Substances having a high specific resistance.

Q. What are conductors?

A. Substances having a low specific resistance. The metals are classed as conductors and the non-metals as non-conductors.

Q. What are insulators?

A. "Insulators" is another name for non-conductors or poor conductors.

Q. What effect does a change of temperature have on the resistance of substances?

TABLE OF RELATIVE RESISTANCES.

(SUBSTANCES ARRANGED IN ORDER OF INCREASING RESISTANCE FOR SAME LENGTH AND SECTIONAL AREA.)

NAME OF METAL.	Resistance in Microhms at 0° Centigrade. 32° Fahr.		Relative Resistance.
	Cubic Centimeter.	Cubic inch.	
Silver, annealed,	1.504	0.5921	1.
Copper, annealed,	1.598	0.6292	1.063
Silver, hard drawn,	1.634	0.6433	1.086
Copper, hard drawn,	1.634	0.6433	1.086
Gold, annealed,	2.058	0.8102	1.369
Gold, hard drawn,	2.094	0.8247	1.393
Aluminum, annealed,	2.912	1.147	1.935
Zinc, pressed,	5.626	2.215	3.741
Platinum, annealed,	9.057	3.565	6.022
Iron, annealed,	9.716	3.825	6.460
Gold-silver alloy (2 ozs. gold, 1 oz. silver), hard or annealed,	10.87	4.281	7.228
Nickel, annealed,	12.47	4.907	8.285
Tin, pressed,	13.21	5.202	8.784
Lead, pressed,	19.63	7.728	13.05
German silver, hard or annealed,	20.93	8.240	13 92
Platinum-silver alloy (1 oz. platinum, 2 ozs. silver), hard or annealed,	24.39	9 603	16.21
Antimony, pressed,	35.50	13.98	23.60
Mercury,	94.32	37.15	62.73
Bismuth, pressed,	131.2	51.65	87.23
Carbon,	. . .	. . .	14.

A. It increases the resistance of metals and diminishes the resistance of non-conductors.

Q. Can you remember about how much a change of temperature of one degree Fahrenheit affects the resistance of metals?

A. It increases the resistance of the common metals roughly about 2 parts in 1000.

Practical Use of Conductors and Insulators.—For carrying electrical energy from the point where it is generated to the point where it is to be used we want to use such material and of such size that the resistance of the circuit does not exceed reasonable limits, although we must be guided by consideration of the first cost. Copper has the lowest specific resistance of the common metals and is generally employed, although if aluminum gets much lower in price than now (30 cts. per pound), it will be a serious competitor to copper. Iron is used only on short telegraph and telephone lines. It is evident that the circuit should be as direct as possible, as the greater its length the greater its resistance, and therefore the greater is the amount of energy lost on the line.

Insulators are used to prevent current from being led off the conductors. For all work except outdoor work, and, indeed, for a large part of that, the conducting wire is covered with one or more layers of some compound of rubber which is a good insulator. The thicker this rubber covering the better its insulating properties, for we have made the path of leakage of current longer by thickening the rubber coating. A further protection is given by suspending the wires at intervals on porcelain or glass or other insulators, so that the wire only comes in contact with its coating, porcelain, or the air, which is also an exceedingly good insulator. To sum up briefly, make the path through which you want the current to flow as short and easy as possible. Make all possible leakage paths as long and narrow as possible.

CURRENT.

Q. What are some of the most notable effects of electric current?

A. It heats the conductors which carry it; it produces around the wire a magnetic field which exerts a force on all magnetic substances placed within the field; it has the power to decompose or electrolyze solutions of many chemical compounds. To these three effects are given the names heating effect, magnetic effect, and electrolytic effect.

Q. Is the heating effect proportional to the strength of current or number of ampères?

A. No; if the ampères are doubled the heating effect is *four times* as great instead of twice as great. With three times as many ampères the heating effect is *nine* times as great.

Q. What is the law, then, which connects the heating effect with the strength of current?

A. The heating effect is proportional to the *square* of the current strength.

Q. How is the heating effect of a certain current affected if the resistance through which it flows is doubled?

A. The heating effect is doubled, it being strictly proportional to the resistance.

Q. Is there any formula which gives the number of heat units produced by a certain current through a certain resistance?

A. Yes; in "Roper's Engineers' Handy-Book," page 670.

Q. Is the heating effect of a current a source of danger?

A. It may be; if wires which carry currents are too small they may be so heated as to set fire to neighboring woodwork. On this account the insurance underwriters have found it necessary to prescribe the minimum sizes which shall be used for various currents. These are published in tables called "Tables of Safe Carrying Capacity of Wires."

Q. Is any practical use made of the heating effect of the electric current?

A. Yes; in electric heaters and cooking devices, and also in the incandescent lamp, where the filament is heated white hot.

Q. Is the magnetic effect of a current proportional to the current strength?

A. Strictly.

Q. Is the electrolytic effect also proportional to the current strength?

A. Yes; doubling the number of ampères will always double the electrolytic effect, tripling the ampères will triple it, and so on.

Q. When, as in Experiment No. 1, a metallic salt is electrolyzed, does the amount of copper deposited bear any definite relation to the current strength?

A. Yes; one ampère will always deposit a definite amount of copper per second.

Q. Does it make any difference what salt of copper is used?

A. Generally speaking, no; but with one or two salts the number of grains of copper deposited per second by one ampère is double what it is with the ordinary salts.

Q. Will one ampère deposit from a silver salt solution the same number of grains per second as with copper?

A. No; one ampère deposits different weights of the various metals per second, the amounts being proportional to the atomic weights of the elements* or to multiples of them.

ELECTRO-MOTIVE FORCE OR ELECTRIC PRESSURE.

Q. In what ways may electric pressure be produced?

A. There are many ways of which these four are the most common:

1. By rubbing together two dissimilar substances, as silk and glass.

2. By heating the point at which two dissimilar metals are joined together.

3. By chemical action, as in Experiment No. 1 with the chemical battery.

4. By moving a magnet relatively to a coil of wire, as in the dynamo, the principle being illustrated in Experiment No. 4.

Q. Which method is the most important?

A. The last; the first two are scarcely used at all in practice. The third is used only where small amounts of power are required.

Q. If there is a difference of electrical pressure existing between two points and these two points be joined by a conductor, what will occur?

* See "Roper's Engineers' Handy-Book," p. 612.

A. An electric current will flow from the point of higher pressure to the other point.

Q. How long will this current continue?

A. As long as there is any difference of pressure between the two points. If the two points are, for example, the terminals of a battery, which by chemical action keeps up a difference of pressure between its terminals, the current would continue until one of the chemicals of the battery, the zinc or solution, is exhausted.

Q. How could you determine if two points were at the same pressure?

A. By connecting a galvanometer between the points. If the needle of the galvanometer was not deflected this would show that no current flowed through it and, therefore, that no difference in electrical pressure existed between the two points to which it was connected.

Q. When an electric pressure exists between two points, is there also any mechanical pressure.

A. Yes; the medium or substance separating the two points is under a mechanical strain which is proportional to the number of volts electrical pressure existing between the two points. If this voltage is very great the substance, be it air, glass, porcelain, or otherwise, is actually cracked and an electric spark passes which tends to relieve the difference of pressure.

OHM'S LAW.

This law, which is the relation existing between current, pressure, and resistance of a circuit, is the most important law in electrical science, and an intelligent application of it will solve most problems which the ordinary engineer will meet. This law is as follows: In an electric circuit the total current (ampères) is equal to the total electric pressure (in volts) divided by the total resistance (in ohms). In shorter form it is expressed by the formula $C = \frac{E}{R}$, where $C =$ current in ampères, $E =$ pressure in volts, and $R =$ resistance in ohms. Several examples will illustrate its use.

Q. In a certain electrical circuit there is an electro-motive force or electrical pressure of 4 volts. The total resistance of the circuit is 2 ohms. How much will be the current?

A. $C = \frac{E}{R} = \frac{4}{2} = 2$ ampères.

Q. What electro-motive force or electrical pressure must be used to force a current of 10 ampères through a circuit whose resistance is 10 ohms?

A. $C = \frac{E}{R}$ or $E = CR = 10 \times 10 = 100$ volts.

Q. If under a pressure or electro-motive force

of 100 volts we get a current flow of 20 ampères, what is the resistance of the circuit?

A. $C = \frac{E}{R}$ or $R = \frac{E}{C} = \frac{100}{20} = 5$ ohms.

When there is more than one electro-motive force acting in a circuit, we must use for the value of E in the above formula the resultant of all the separate electro-motive forces acting. When there are several resistances in a circuit their joint resistance must be used.

Q. Suppose we have two batteries, one giving 2 volts and the other 1 volt, their plates being zinc and carbon, but different solutions being used in each. Connect the zinc of one to the carbon of the other, and then connect from A to B a piece of wire having a resistance of, say, 10 ohms, as shown in the sketch. When connected in this way the electro-motive forces are added, and the total electro-motive force is $2 + 1$, or 3 volts. The batteries themselves have some resistance, and also the lead wires $A\ C$ and $B\ D$. Suppose that the resistance of one battery is 4 ohms and the other 2 ohms, the resistance of $A\ C$ and $B\ D$ each 1 ohm. Then the total resistance of the circuit is $10 + 1 + 2 + 4 + 1 = 18$ ohms. What will be the current?

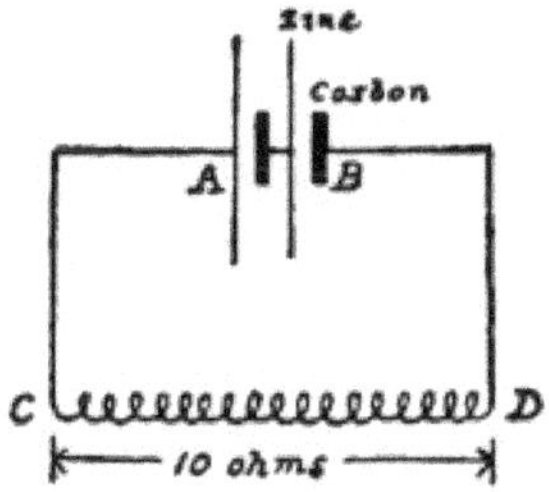

A. The current will be $\frac{\text{resultant } E}{\text{Total } R} = \frac{3}{18} = \frac{1}{6}$ ampère.

Q. Suppose that one of the batteries was reversed so that the two zincs are connected together as in the sketch?

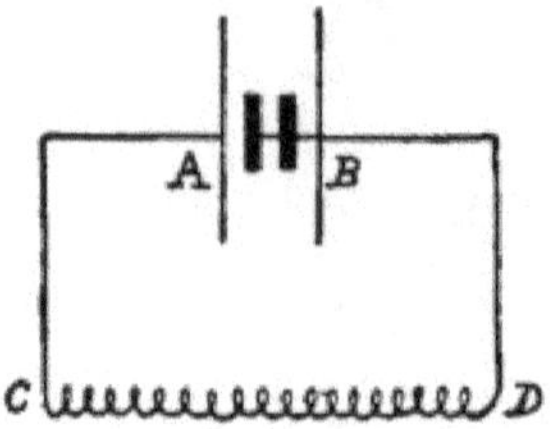

A. The batteries now oppose each other and the resultant or effective electro-motive force is 2—1, or 1 volt. The resistance of the circuit is, as before, 18 ohms, and the current will be $\frac{1}{18}$ ampère.

Calculation of Current in Divided Circuits.—Suppose that the battery has an electro-motive force of 2 volts, that its resistance is $\frac{1}{3}$ ohm, that the resistance of the lead wire *A B* is 3 ohms, and that between *C* and *B* we have two paths of resistance 10 and 20 ohms each.

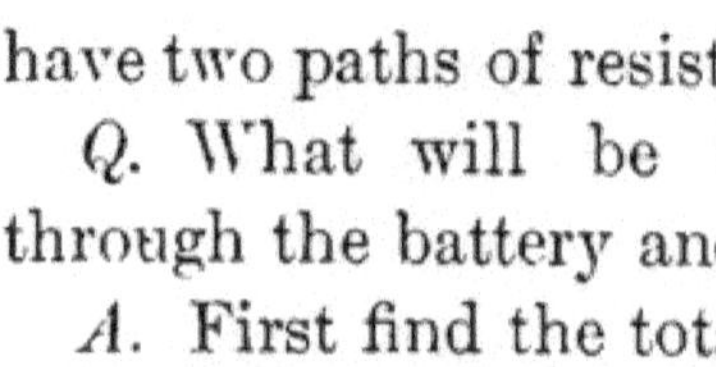

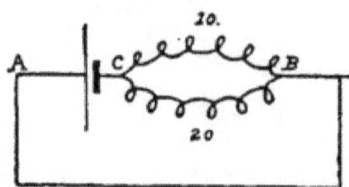

Q. What will be the total current flowing through the battery and through *A B*?

A. First find the total resistance of the circuit. The joint resistance between the points *B* and *E* is, as previously shown under "Resistance," equal to $\frac{10 \times 20}{10 + 20} = \frac{200}{30} = 6\frac{2}{3}$ ohms. The total resistance of the circuit is therefore $6\frac{2}{3} + \frac{1}{3} + 3$, or 10 ohms. The current is equal to $\frac{E}{P} = \frac{2}{10} = .2$ ampère.

Q. What part of the current flows through each branch?

A. Obviously the greater part of the current will flow through the branch having the smaller resistance. $\frac{10}{30}$ or $\frac{1}{3}$ ampère will flow through the 20 ohms branch, and $\frac{20}{30}$ or $\frac{2}{3}$ ampère will flow through the other branch.

PRACTICAL APPROXIMATION.—If the resistance of batteries or generator and the leads is small compared to that of the main resistance in circuit, we may neglect them, using for R in the formula the resistance of the external circuit. This is generally the case in electric lighting circuits, where the resistance of the generator will rarely exceed one-hundredth of an ohm, and where the resistance of the line wires will usually be less than one-twentieth of the joint resistance of the lamps.

Example.—*Q.* On a 110-volt circuit, what is the current (total) when one sixteen-candle-power lamp of 220 ohms' resistance is turned on?

A. $E = 110$, R is practically 220 ohms. The current $= \frac{110}{220} = \frac{1}{2}$ ampère.

Q. What is the current (total) when two lamps are turned on?

A. The joint resistance of two similar lamps is $\frac{220 \times 220}{220 + 220} = \frac{220 \times 220}{2 + 220} = 110$ ohms, or *half* that of one lamp. The total current $= \frac{110}{110} = 1$ ampère. The current through each lamp is the same, and is $\frac{1}{2}$ ampère as before.

With three lamps turned on the joint resistance is one-third of 220, or $73\frac{1}{3}$, and the total current is $\frac{110}{73\frac{1}{3}} = 1\frac{1}{2}$ ampères, and the current through each lamp is still $\frac{1}{2}$ ampère. Turning on one lamp then adds $\frac{1}{2}$ ampère to the total current. The lamps are connected in multiple as shown in the figure.

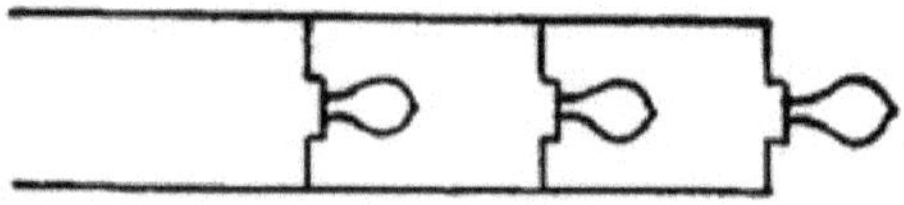

The Use of Alternating Currents complicates the calculation of current, pressure, and resistance by Ohm's law, and the method of making such calculations is outside of the scope of this book, inasmuch as the ordinary engineer would rarely be called upon to do so.

ELECTRICAL MEASUREMENT.

Q. What are the electrical quantities which the engineer is called upon to measure?

A. Current, electro-motive force, resistance, and power.

Q. What instruments are necessary?

A. For direct-current circuits, an ammeter and voltmeter of proper range.

Q. How are the Weston ammeters constructed?

A. They consist of a fixed permanent magnet of horse-shoe form, between the poles of which is pivoted a coil of fine wire which carries the needle. When the coil is connected so that a current flows through the coil, it tends to turn so as to include the maximum number of lines of force due to the magnet. This motion is resisted by a pair of springs resembling the hair spring of a watch.

In the instruments for measuring currents of more than an ampère, only a known fraction of the current passes through the coil, the balance passing through a conductor placed in parallel with the coil.

Q. Suppose we have a circuit similar to that in the sketch and we desire to measure the current taken by four lamps. How would you proceed?

A. If these are 16 candle-power (16 c. p.)

lamps on a 110-volt circuit, we know that they will take, roughly, ½ ampère each. Therefore to measure accurately their current we need an ammeter intended to measure small currents. Connect its terminals to two points on the circuit as *C* and *D* by wires, as shown by dotted lines. Then cut the circuit between *C* and *D*. The total current will now flow around through the ammeter and the reading of the needles will, if the instrument is correct, give the current in ampères.

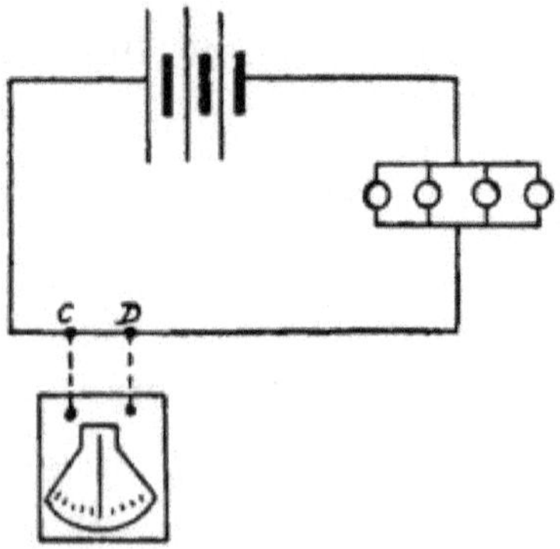

Notice that one terminal is marked + and the other —. If the instrument is not connected properly, the needle will move, or try to move, to the left of the scale. In this event reverse the wire connections from the points *C* and *D* to the instrument. Such an instrument tells the polarity of the circuit—that is, which is the higher pressure and which the lower pressure side. When the + binding-post is connected to the higher pressure side of the circuit the needle deflects in the proper direction.

Q. Suppose we have no ammeter of proper range available, but we have a resistance whose value we know and which will carry the current to be measured without much heating?

A. In this case with the aid of the voltmeter we can measure current. Suppose we have a resistance which we know is 1 ohm and a portable voltmeter with an additional scale reading from 0 to 15 volts, and we want to make the current-measurement just described. Put the resistance in between *C* and *D* and connect the voltmeter terminals to the ends of the resistance. Suppose the reading of the voltmeter was 2.3 volts. The current through the resistance is by Ohm's law equal to the electrical pressure or electro-motive force between its terminals divided by the resistance, or 2.3 ÷ 1, which is 2.3 ampères. This is the method used in the Weston switchboard instruments, a resistance of known value being placed in the main circuit of the dynamo and two leads taken off from its terminals and run to a voltmeter.

Q. How would you measure the electrical pressure between two points?

A. I would connect the terminals of a voltmeter, one to each of the points.

Q. Suppose the voltage between the points is greater than the range of the voltmeter. For example, suppose you wish to measure a voltage which you know is about 220, but have an instrument which reads only to 150 volts, what is the method?

A. Connect between the two points *A* and *B*, whose voltage is wanted, two 110-volt lamps in series. Then make the connections shown by the solid lines and read. Change the connections to the dotted positions and read again. The sum of the two readings will be the voltage between *A* and *B*.

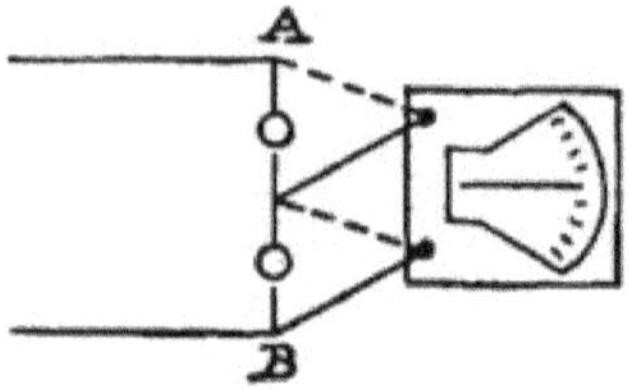

Q. Is there any other method?

A. Yes; in the other method it is necessary to have a known resistance, to place it in series with the voltmeter, and also to know the resistance of the voltmeter. This last is usually given on the box containing the instrument. A resistance just equal to that of the instrument doubles its range. In general, to get the value of the reading of a voltmeter when a resistance has been put in series with it, multiply its reading by the sum of the resistance of the instrument and the auxiliary resistance, and divide the product by the resistance of the instrument.

Q. How would you measure a resistance; for instance, the resistance of a coil of wire?

A. If I had an ammeter and voltmeter of proper range I would put the ammeter in series with the coil and would connect the voltmeter to its terminals. Then I would send a current from a battery or dynamo through the coil and take the readings of the ammeter and voltmeter. By Ohm's law current $= \frac{\text{voltage}}{\text{resistance}}$ or resistance $= \frac{\text{voltage}}{\text{current}}$.

Q. What do you mean by instruments of proper range in this case?

A. The ammeter must be suitable for measuring the largest current which the coil can carry without overheating, and the voltmeter must be such that the voltage at the terminals of the coil will give a deflection of the need large enough to be readable with accuracy.

Q. Is there any other method of measuring resistance?

A. Several. One of the most valuable, since it needs only a voltmeter of known resistance and some form of current generator, is known as the *Voltmeter Method.* This method requires two readings of the instrument. For the first reading the instrument is connected to the terminals of the

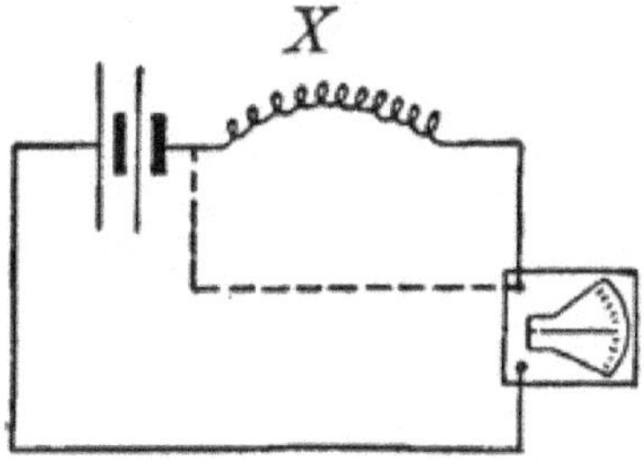

current-generator. For the second reading the unknown resistance is put in series with the voltmeter and then the two connected to the generator. In the figure X is the unknown resistance, and for the first reading the connection shown by the dotted line is made. For the second reading the connection is as shown by the solid lines. To calculate the resistance from the readings divide the first reading by the second, then multiply the quotient by the resistance of the voltmeter, and from the product subtract the resistance of the voltmeter.

Q. Which of these methods would you use for low resistances of, say, less than 100 ohms?

A. The first method.

Q. Which for high resistances, such as insulation tests?

A. The voltmeter method.

Q. How would you connect for a test of the insulation of the armature coils of a dynamo, from the frame?

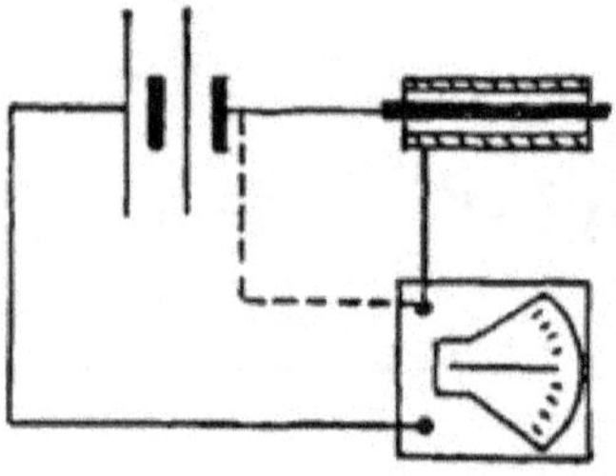

A. As in the figure, the heavy black line representing a commutator segment, and the cross-hatched portion representing the frame. The white space between, of course, represents the insulating material.

Q. How would you measure the power used in any part of a circuit, as, for example, in a lamp?

A. Power being the product of volts by ampères (in direct-current circuits), I would connect an ammeter in series with the lamp and a voltmeter to its terminals, and would multiply their readings together, thus obtaining the number of watts.

Q. Suppose you wished to get the horse-power?

A. I would divide the number of watts by 746.

ELECTRIC BATTERIES.

Q. What two kinds of electric generators are in most common use?

A. The chemical generators, or batteries, and the magneto-electric generators, or dynamos.

Q. In what cases are batteries used?

A. When the amount of power to be supplied is small, as for bells, time clocks, telegraphs, telephones, surgical lamps, dental engines, etc., and in some cases in which the introduction of the engine which would be needed to drive a dynamo would be objectionable.

Q. Why are batteries not used when large amounts of power are required?

A. On account of the expense of the chemicals used. Zinc is in nearly all batteries the fuel, and since the energy produced by burning one pound of it is only one-sixth that produced by one pound of coal, and, moreover, since the cost of zinc is about sixty times that of coal, it is much cheaper to generate electric power by means of coal rather than by means of zinc.

Q. What are secondary or storage batteries?

A. Those whose chemical actions may be reversed by sending an electric current (from some outside source) through them in the opposite direction to the current which they have produced. Thereby they are restored to the original condition which existed before they were used to produce electric current.

Q. Do they store electricity?

A. Not at all. They store up energy in the form of chemical energy, which at any time may be changed into electrical energy by connecting the terminals of the battery together by some conductor.

Q. What are primary batteries?

A. Those whose chemical actions cannot be reversed by passing an electric current through them in the reverse direction.

Q. Give an example of a reversible cell.

A. The Daniell cell.

Q. Is it used as a storage or as a primary battery?

A. As a primary; others being better adapted for use as secondaries.

Q. Into what two classes may primary cells be divided?

A. Open-circuit cells and closed-circuit cells.

Q. What is an open-circuit cell?

A. A cell suitable for use on circuits that are normally open, being closed only at the moment when work is to be done; as, for example, bell circuits, gas-lighting circuits, time systems, watch-clock systems, etc.

Q. What kind of a cell is generally employed for such work?

A. A cell known as the Leclanché, having a zinc plate for one pole, a carbon plate for the other pole, and the two immersed in a solution of sal-ammoniac.

Q. What is the voltage furnished by such a cell and what is the resistance of the ordinary size cell?

A. About $1\frac{1}{2}$ volts and from $\frac{1}{10}$ to $\frac{5}{10}$ ohm resistance.

Q. Why is not this cell suitable for closed circuit work?

A. Because when a circuit is closed hydrogen particles begin to collect on the carbon plate, and these cut down the voltage and at the same time increase the resistance of the cell.

Q. If the circuit of the cell is opened do these disappear?

A. Yes; in a few minutes.

Q. Is there any way of lessening the trouble caused by the collection of hydrogen particles?

A. Yes; by using a porous carbon and by putting next to the carbon a slab of some strong oxidizing agent like manganese binoxide. In the best forms of cell the carbon is made in the form of a thin, hollow cylinder, and the manganese in powdered form is placed inside.

Q. What is the effect of the manganese binoxide?

A. It gives up a part of its oxygen, which attacks the hydrogen particles and forms, with them, water.

Q. Why are some zincs made in the form of a hollow cylinder extending around the carbon?

A. To diminish the resistance of the cell. The greater the surface of the plates and the nearer they are together, the less is the resistance of the cell.

Q. What cell is largely used for closed circuit work?

A. Some form of the Daniell cell. In its original form it consisted of a zinc plate in sulphuric acid on one side of a porous wall and a copper plate in a solution of copper sulphate on the other side.

Q. What is the gravity cell?

A. A form of Daniell in which the different specific gravities of the liquids are used to keep the liquids from mixing without the use of a porous cup.

Q. What is the voltage and resistance of a Daniell cell?

A. The voltage is about 1 volt. The resistance of the ordinary size gravity is in the vicinity of 4 ohms.

Q. What other cell is largely used and for what class of work?

A. The bichromate cell; for small motors and cautery work, where a strong current is needed for a few minutes. It consists of zinc and carbon plates immersed in chromic acid.

Q. What is the voltage of these cells and their resistance?

A. About 2 volts. Their resistance varies, of course, with their size, that of the smaller sizes being only a fraction of an ohm.

Q. What are the two chief objections to this cell?

A. The fumes produced and the eating of zinc even when the circuit is open.

Q. What is done to lessen the latter objection?

A. The cell is arranged so that the zinc plate can be easily raised out of the solution when the circuit is open.

Q. What are dry cells?

A. Cells in which the solution has been reduced to a pasty condition.

Q. What are their advantages?

A. Their greater portability; on the other hand, their resistance is higher, and they *polarize* more readily.

Q. What do you mean by polarization?

A. The collecting of hydrogen particles previously mentioned.

DYNAMOS.

Q. For what is a dynamo used?

A. To change mechanical energy into electrical energy.

Q. The dynamo as well as the battery are sometimes likened to an electrical pump. In what respect do they resemble a pump?

A. They may be considered as raising electricity from a low level to a high level, just as a pump raises water.

Q. Of what does a dynamo consist?

A. Of a magnet and a coil of wire moving relatively to each other. Generally, the magnet is fixed and the coil rotates between its poles. A difference of electric pressure is set up between the two ends of the coil, and if these ends are connected together a current will flow.

Q. Upon what does the amount of electrical pressure depend?

A. It is proportional to the rate of change in the number of lines of force enclosed by the coil. It is, therefore, increased by increasing the strength of the magnet, the speed of revolution, or the number of turns of wire in the coil.

Q. With such a simple dynamo, is the direction and strength of current uniform?

A. No; the current can best be represented by plotting its values at different moments, as in the figure. Here distances to the right along the horizontal line represent time. Distances above or below the line represent the strength of current at different times. The curve shows the variation of current during three complete revolutions of the coil. It is evident from this curve that the strength of current is always changing and that it changes direction twice in each revolution. *

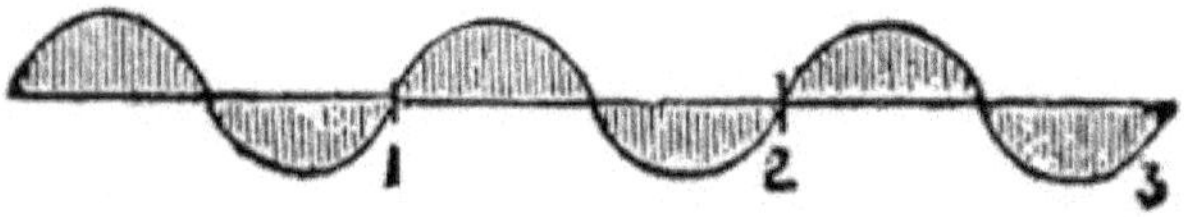

Q. What is such a current called?

A. An alternating current.

Q. Can it be used for practical purposes?

A. Yes; for lighting and for small motors.

Q. How is the current *rectified* or made continuous in direction in the circuit where it is to be used?

A. By the commutator, a purely mechanical device which changes the connection between the ends of the coil and the external circuit just at the moment that the direction of the current in the coil is reversed.

* See also "Roper's Engineers' Handy-Book," page 689.

Q. What is a rectified current called?

A. A direct current.

Q. For what purposes is it employed?

A. For nearly all isolated lighting plants, for operating most arc lights, for driving motors, and for charging storage batteries.

Q. What is the moving coil called?

A. The armature.

Q. How does it differ in practice from the ideal simple dynamo?

A. The armature is made up of a large number of coils wound on an iron core. The larger number of coils give greater uniformity to the strength of current and diminishes the sparking at the commutatôr. The iron core is used to keep as many as possible of the lines of force produced by the magnet in the space in which the armature is moving, thus making the electrical pressure higher than would be the case without the iron core.

Q. How is the iron core made?

A. Of thin circular disks held together by bolts and attached to the armature shaft by a sort of spider.

Q. What two classes of armatures are there?

A. The Gramme ring and the drum-wound.*

Q. What is the reason of making the core out of disks instead of solid metal?

*See "Roper's Engineers' Handy-Book," page 691.

A. To diminish the heating of the core by useless currents set up in the core.

Q. Are the disks separated from each other in any way?

A. They are insulated from each other by enamel or by thin sheets of varnished paper.

Q. Is the field magnet of the dynamo a permanent or electro-magnet?

A. An electro-magnet excited by coils carrying either a part or all of the current supplied by the dynamo.

Q. What is a series machine?

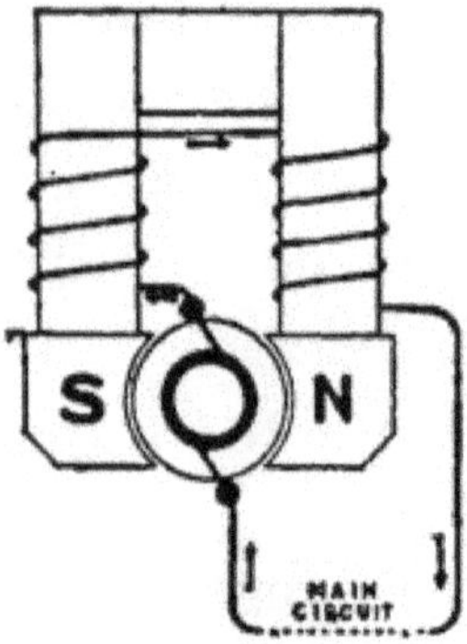

SERIES MACHINE.

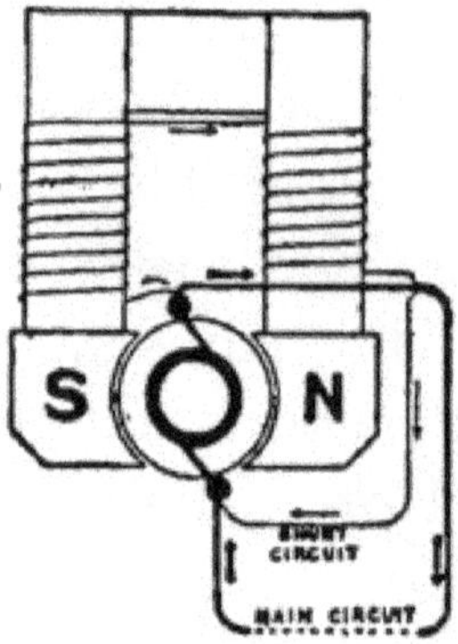

SHUNT MACHINE.

A. A dynamo in which the field-magnet coils carry all the current produced by the machine—that is, the current flows around the field-magnet coils before going to the external circuit.

Q. What is a shunt dynamo?

A. One in which only a fraction of the current is had around the field-magnet coils.

Q. What is a compound dynamo?

A. A combination of shunt and series.

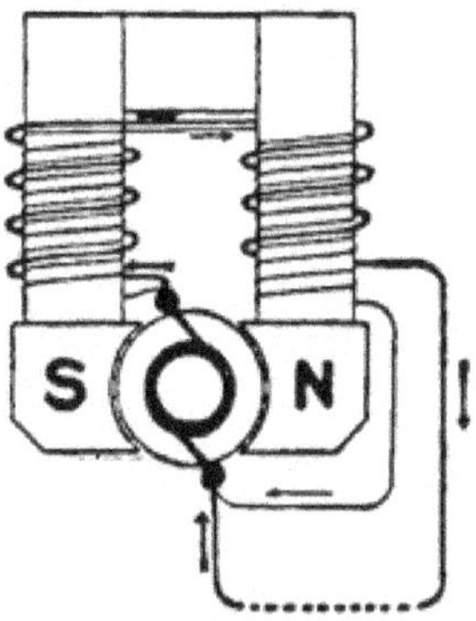

COMPOUND MACHINE.

Q. What are the purposes for which a series dynamo is used?

A. A series dynamo tends to produce a current of constant strength whatever load may be thrown on it. It is therefore used for constant-current circuits such as street arc lighting.

Q. When is the shunt machine used?

A. When a machine is desired which will supply constant *pressure* at all loads.

Q. Does a shunt machine do this?

A. Quite well, but if the closest regulation for constant pressure is desired a compound machine is used.

Q. What is an over-compounded machine?

A. One which, instead of maintaining the pressure constant as the load increases, will raise the pressure a few volts proportionally to the amount of load.

Q. What is the advantage of this?

A. There are two advantages. One is to make

up for a slight lowering of speed in the engine, which takes place as the load increases. The other is to make up for the loss in pressure owing to the resistance of the external circuit wires, which loss is proportional to the load which they carry.

Q. How can the pressure furnished by a shunt or compound dynamo be varied?

A. An adjustable resistance called a rheostat is connected in series with the shunt-field coils; by turning the arm of the rheostat in one direction more resistance is thrown into this circuit and the current flowing around the coils is diminished. This cuts down the number of lines of force produced by the field magnet, and therefore the pressure furnished by the machine is lowered. Moving the rheostat arm in the other direction raises the pressure by cutting out resistance.

Q. What are the brushes?

A. The brushes are pieces of copper or carbon resting on the commutator and serving to take current from the commutator to the external circuit.

Q. In order to secure freedom from sparking what care must be exercised in setting the brushes?

A. The brushes must be opposite each other, and must fit the surface of the commutator properly. The rocker arm carrying them must be turned into the position of least sparking.

DISTRIBUTION OF ELECTRICAL ENERGY.

The production and distribution of electrical energy are very much like a small water-system, where water is pumped from a tank to a high reservoir, taken from the reservoir through pipes to the place where it is to be used, and *after use* led back to the tank to be again pumped up and again used. The generator, or dynamo, driven by a steam engine, gas engine, or water-wheel, corresponds to the pump. The distributing-pipes in the water-system are replaced by copper wires for the electrical system. The high-pressure reservoir and low-pressure tank are replaced by the switchboard bus bars, one of which is a high-pressure and the other a low-pressure bar. The high-pressure bar is also called the positive or plus (+) bar, and the other the negative or minus (—) bar. They are each copper bars mounted on the marble or slate of which the switchboard is made, and are called *bus bars*, or omnibus bars, from the fact that all the current is carried by them. The valves of the water-system are replaced by switches, the water-meters by ammeters, and pressure-gauges by voltmeters. Some devices which are used in electrical distribution have nothing similar to them in water-systems, but the general similarity is of great assistance in understanding electrical distribution.

Q. What is a switchboard?

A. One or more slate or marble slabs mounted on an iron or wooden framework and containing the various devices for controlling the electric distribution system.

Q. What are the principal devices to be found on the switchboard?

A. 1. A voltmeter to measure electric pressure. This is generally furnished with a switch by which it may be connected to the terminals of any generator or to the *bus bars.*

2. An ammeter for each generator to measure the current which it furnishes.

3. A rheostat for each generator placed in series with its shunt-field coils and controlling the pressure furnished by it.

4. A device for each machine, such that if owing to any trouble a current greater than the maximum for which the machine is designed flows through the machine, it is automatically disconnected from the circuit. This device may be a fuse or a circuit breaker.

5. A device called a *ground detector*, for showing when the conductors in the system are by accident brought into electrical connection with the earth;

that is to say, with gas- or steam- or water-pipes which are imbedded in the earth.

6. Switches for disconnecting the generators from the bus bars.

7. Switches for disconnecting from the bus bars the distribution circuits.

8. A device (either fuse or circuit breaker) for protecting each distribution circuit from having too much current flow over it.

Q. What are fuses?

A. Strips of an alloy, generally of tin and lead, of such size that they will melt and interrupt the circuit when a current in excess of a certain amount flows through them.

Q. What are circuit breakers?

A. Switches so arranged that they open automatically when the current flowing through them exceeds a certain value.*

Q. Why are circuit breakers used in preference to the much cheaper fuses?

A. Because in large sizes fuses are very uncertain in their action; a fuse designed to melt at 500 ampères, for example, being liable to melt with a current of 400 or 600 ampères.

Q. How is a simple form of ground detector made, and how does it operate on a circuit, say, whose pressure is about 110 volts?

*See "Roper's Engineers' Handy-Book," page 705.

A. The ground detector consists of two 110-volt lamps connected in series with each other and across or between the bus bars. The junction between the two lamps is connected to a convenient water-pipe. So long as the insulation of the circuit is all right the two lights burn alike equally dim, since they are designed for 110 volts at their terminals and they have only 55 volts under the circumstances. But suppose any point on the circuit, as *P*, is purposely or accidentally connected to earth, then the left-hand light will burn bright while the right-hand one will burn exceedingly dim, or perhaps not at all. The reason is that the grounding of the point *P* has put it in electrical connection with the point *A* through a very low resistance. The current through the right-hand lamp is, therefore, diminished, its terminals being short-circuited. The left-hand lamp will have practically 110 volts between its terminals, since the joint-resistance of the right-hand lamp and the other path from *A* to *P* is exceedingly small, and hence the pressure used up being also exceedingly small. If the point *P* were on the other side of the circuit, the right-hand lamp would burn brightly and the left-hand one very dimly.

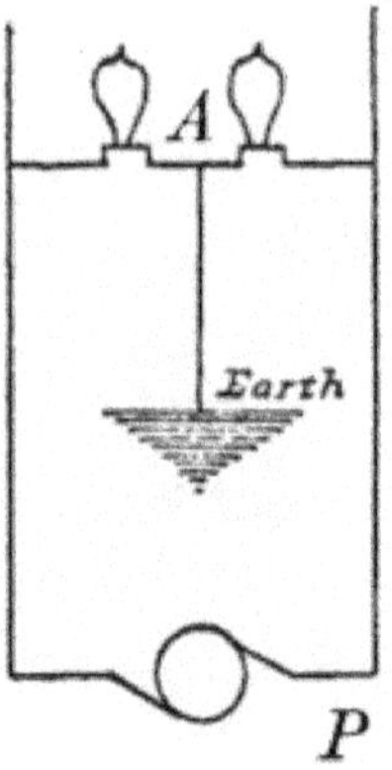

Q. How would you find the location of the ground?

A. By opening the switches one by one till one is found which on being opened relieves the ground. This tells on which feeder the ground exists. Then the circuit is examined in detail by means of a magneto-bell, it being split up into sections by throwing open local switches, taking fuses out of local distribution boards, and disconnecting at fixtures.

Q. May any number of dynamos be connected in multiple so as to feed on the same pair of bus bars?

A. Any number of shunt machines of the same voltage may be so used.

Q. Cannot compound machines be so connected?

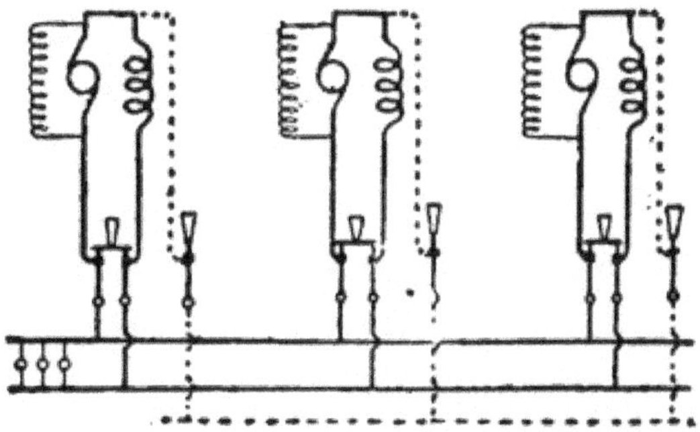

A. Not without a connection called the *equalizer* shown by the dotted line in the cut.

Q. Suppose you have one machine feeding the bus bars and desire to connect up with it machine No. 2, how would you proceed?

A. First start up the engine of No. 2 and turn its rheostat till its pressure is the same as that of the bus bars or perhaps one-half volt higher. Then close the single-pole switch in the equalizer circuit, shown dotted, and finally close the machine's double-pole switch which connects it to the bus bars. Its ammeter reading will then increase, and the rheostat handles of the two machines are moved till the ammeters read alike (if the machines are the same size) and the voltage of the bus bars is correct.

Q. Is any different arrangement of switches ever employed?

A. Yes; instead of a two-pole switch in the dynamo leads and a single-pole switch in the equalizer lead, a three-pole switch is frequently employed. In this case the middle blade is used for the equalizer wire, and is so adjusted that it closes the equalizer circuit just before the other two blades close their circuits.

SYSTEMS OF DISTRIBUTION.

Q. What are the two principal systems of electrical distribution?

A. The series system and the parallel system.

Q. What is the difference between the two systems?

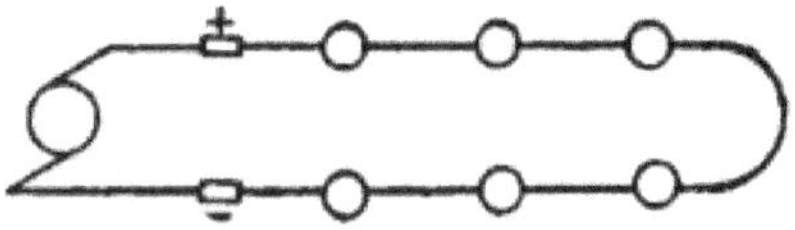

A. In the series system the entire current flows successively through each lamp. In the parallel

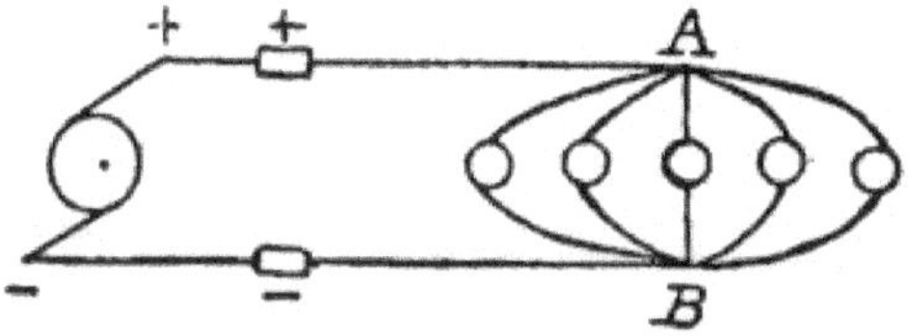

system the current from the dynamo is divided, a part flowing through each lamp. Afterward these separate currents unite and flow back to the dynamo.

Q. What is necessary, on a series system, to make the lighting successful?

A. It must be a constant-current system—that is, cutting out lamps or throwing more on must not change the value of the current.

Q. How is this accomplished?

A. By an automatic regulator on the machine which increases its voltage if lamps are thrown on, and diminishes it if lamps are cut out.

Q. How are lamps cut out on this system?

A. By short-circuiting them—that is, by providing another path for the current to flow other than the path through the lamp mechanism and carbons.

Q. What is necessary in a parallel system?

A. It must be a constant-potential or constant-pressure system.

Q. How are lamps cut out on this system?

A. By interrupting the branch circuit in which the lamp is connected.

Q. In the parallel system, why does cutting out one lamp not affect others?

A. Because it does not change the current flowing through each of the others. The current through any lamp depends on two things only,—the pressure and the resistance of the lamp. Turning out a lamp in nowise affects the resistance of other lamps and only affects the pressure at the terminals to a very slight degree; therefore the current flowing through the lamp is practically the same as it was before the other lamp was turned off.

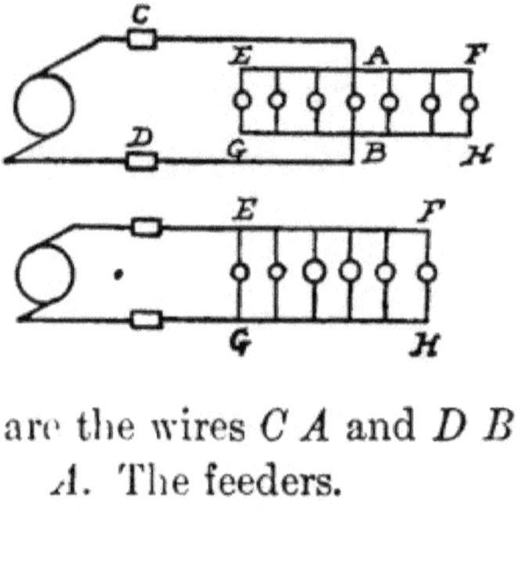

Q. In the cut, what are the wires *C A* and *D B* called?

A. The feeders.

Q. And the wires *E F* and *G H*?

A. The mains.

Q. And from *F* to the lamp and *H* to the lamp?

A. Branches.

Q. What is the Edison three-wire system?

A. Two 110-volt machines are connected in series and the middle or neutral wire is connected to their junction. When the same number of lamps are burning on each side of the neutral wire there is no current flowing through the neutral and the same current flows through each machine. When No. 4 is turned out, for example, the lower

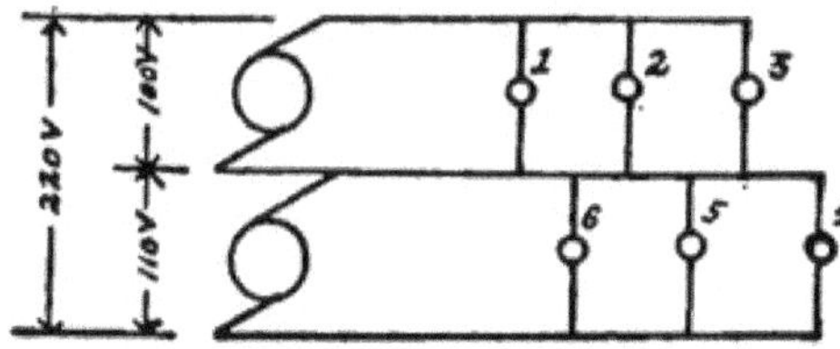

machine supplies only the current necessary for lamps 5 and 6, while the upper continues to supply the same as before, the current for one lamp returning to the upper machine over the neutral. If all lamps on one side were turned out, the machine on that side would furnish no current, and the other machine would continue to work as before.

Q. What is the advantage of this system?

A. It is a 220-volt system and therefore requires

much smaller wires to transmit a given amount of energy with a given loss, without increasing the voltage of the lamps.

Q. How much is the gain in size of wire used?

A. The two outside wires are just one-quarter as large as they would be with a 110-volt two-wire system. If the neutral is made of the same size, the three-wire system requires $\frac{3}{8}$ as much copper as the two-wire system, using the same voltage lamps in both cases.

TABLE

SHOWING GAIN BY USING HIGH PRESSURES, THE SAME SIZE WIRES BEING USED FOR EACH CASE.

Power transmitted in watts. $C \times E$	Volts at which transmitted. E	Corresponding number of ampères. C	Power lost in watts. $C^2 R$	Volts drop in line. $C R$	Per cent. power lost. $C^2 R \div 1100$	Per cent. volts lost. $C R \div E$
1100	110	10	100	10	11.	9.9
1100	220	5	25	5	2.75	2.27
1100	550	2	4	2	.0227	.363
1100	1100	1	1	1	.0009	.091

Q. If in one case, to transmit a certain power, we use 110 volts' pressure and in another case 1100 volts, what will be the relative amount of copper used on the line?

A. With 1100 volts' pressure we shall need only $\frac{1}{100}$th as much copper as with 110 volts.

Q. What disadvantages have high pressures?

A. Greater difficulty in insulating the lines and danger to human life.

Q. In proportioning the size of electrical conductors, what two requirements must be met?

A. The wire must be large enough to transmit the energy without losing more than a prescribed per cent., and the wire must further be large enough so that the current will not heat it more than is allowed by the insurance regulations.

INSURANCE RULES FOR CARRYING-CAPACITY OF WIRES.

B. & S. gauge.	National Electric Light Association.	National Board of Fire Underwriters.		Assoc. Factory Mutual Ins. Co.	English Board of Trade.
		Concealed.	Open work.		
0000	175	218	312	175	
000	145	181	262	145	
00	120	150	220	120	105
0	100	125	185	100	83
1	95	105	156	85	66
2	70	88	131	70	52
3	60	75	110	60	41
4	50	63	92	50	33
5	45	53	77	45	26
6	35	45	65	35	21
7	30	..	..	30	16
8	25	33	46	25	13
10	20	25	32	20	8
12	15	17	23	15	5
14	10	12	16	10	3
16	5	6	8	5	2
18	..	3	5	3	1

Q. What is the loss of pressure allowable on conductors?

A. See "Roper's Engineers' Handy-Book," pp. 714–717.

Q. The distance between the switchboard and a group of ten 16 c. p. lamps is 100 feet. What size wire must be used so that the loss of pressure on the wire between switchboard and lamp is only one-half of one per cent., the voltage of the dynamo being 110?

A. 1. One-half of one per cent. of 110 is .55 volt, the allowable loss of pressure.

2. The current for ten lamps is 5 ampères.

3. By Ohm's law $C = \frac{E}{R}$ or $R = \frac{E}{C}$. $R = \frac{.55}{5}$ $= .11$ ohm—that is, the wire must be of such size that the total length of it, 200 feet, has a resistance not exceeding .11 ohm; 1000 feet of this size wire would have a resistance $\frac{.11 \times 1000}{200}$ $= .55$.

4. Looking in the wire tables we see that No. 7 wire, having a resistance of .491 ohm at 60° Fahr. fulfils the requirement.

5. Looking in the table of safe carrying capacities on the preceding page, we find that according to the National Board of Fire Underwriters' rules a No. 7 wire will carry a much greater current

PROPERTIES OF COPPER WIRE.

ENGLISH SYSTEM—BROWN & SHARPE GAUGE.

Numbers.	Diameters in mils.	Areas in circular mils. C. M. = d².	Weights.		Resistances per 1000 feet in International ohms.	
			1000 feet.	Mile.	At 60° F.	At 75° F.
0000	460.	211600.	641.	3382.	.04811	.04966
000	410.	168100.	509.	2687.	.06056	.06251
00	365.	133225.	403.	2129.	.07642	.07887
0	325.	105625.	320.	1688.	.09639	.09948
1	289.	83521.	253.	1335.	.1219	.1258
2	258.	66564.	202.	1064.	.1529	.1579
3	229.	52441.	159.	838.	.1941	.2004
4	204.	41616.	126.	665.	.2446	.2525
5	182.	33124.	100.	529.	.3074	.3172
6	162.	26244.	79.	419.	.3879	.4004
7	144.	20736.	63.	331.	.491	.5067
8	128.	16384.	50.	262.	.6214	.6413
9	114.	12996.	39.	208.	.7834	.8085
10	102.	10404.	32.	166.	.9785	1.01
11	91.	8281.	25.	132.	1.229	1.269
12	81.	6561.	20.	105.	1.552	1.601
13	72.	5184.	15.7	83.	1.964	2.027
14	64.	4096.	12.4	65.	2.485	2.565
15	57.	3249.	9.8	52.	3.133	3.234
16	51.	2601.	7.9	42.	3.914	4.04
17	45.	2025.	6.1	32.	5.028	5.189
18	40.	1600.	4.8	25.6	6.363	6.567
19	36.	1296.	3.9	20.7	7.855	8.108
20	32.	1024.	3.1	16.4	9.942	10.26
21	28.5	812.3	2.5	13.	12.53	12.94
22	25.3	640.1	1.9	10.2	15.9	16.41
23	22.6	510.8	1.5	8.2	19.93	20.57
24	20.1	404.	1.2	6.5	25.2	26.01
25	17.9	320.4	.97	5.1	31.77	32.79
26	15.9	252.8	.77	4.	40.27	41.56
27	14.2	201.6	.61	3.2	50.49	52.11
28	12.6	158.8	.48	2.5	64.13	66.18
29	11.3	127.7	.39	2.	79.73	82.29
30	10.	100.	.31	1.6	101.8	105.1
31	8.9	79.2	.24	1.27	128.5	132.7

There are two points in this table which will be found easy to remember and very convenient in practice—namely, that the resistance of 1000 feet of No. 10 is almost exactly 1 ohm at 75° F., and that a change of three sizes either halves or doubles the resistance, according as we go up or down the table.

than 5 ampères, so that a No. 7 wire is suitable for the requirements.

Q. What is a mil?

A. One-thousandth of an inch.

Q. What are the circular mils in a wire?

A. The square of the diameter in mils.

Q. What relation do the circular mileages of two wires bear to their resistances?

A. Their resistances are inversely proportional to their circular mileages.

Q. A No. 2 wire, No. 4 wire, and No. 6 wire are connected in multiple; to what size wire will their joint resistance be equal?

A. The sum of their circular mileages is,— $66,564 + 41,616 + 26,244 = 134,424$, and this is nearly the circular mileage of a No. 2/0 wire to which the three wires will be practically equivalent.

WIRING AND APPLIANCES.

Q. What two classes of wiring are there?

A. Open or exposed work and concealed work.

Q. In open work, what varieties are there?

A. Porcelain work, where the wires are carried on porcelain knobs, and molding work, where the wires are carried in a grooved molding provided with a cap to hide them from view.

Q. What are the varieties of concealed work?

A. Porcelain work and conduit work.

Q. What is the nature of conduit work?

A. A system of tubes or pipes is first installed into which the wires are afterward drawn in.

Q. What are the fundamental requisites for a conduit?

A. It should be strong enough to protect the wires from all accidents such as hammering, jarring, nails, etc., and it should not be attacked by cement, plaster, or moisture. Moreover, it should have a smooth inside surface, so that the insulation of the wires may not be injured by the process of drawing them in.

Q. What kind of conduits meet these requirements?

A. An iron or steel tube like a gas-pipe has sufficient strength. If properly painted or enameled it is not affected by cement, plaster, or moisture. To secure smoothness a special pipe must be made, with this end in view; or, as in some conduits, a lining of wood or some compound of a bituminous nature may be employed.

Q. How many wires are placed in one tube?

A. Two in the two-wire system or three in the three-wire system, except sometimes in the case of large-sized feeders where it is not possible to draw two in. Where alternating currents are to be used both the wires of a circuit *must* be in the same tube to avoid an excessive loss of pressure.

Q. What is a cut-out, and when is it used?

A. A cut-out is the name given to a combination of fuse blocks, studs, and screws and convenient terminals for fastening wires. These parts are mounted on some insulator, as slate, marble, or porcelain. A cut-out with fuse is used at every point in a circuit where the size of wire is changed.

Q. Why is this?

A. So that the fuse may protect the smaller wire from an excess of current.

Q. What is a switch?

A. A convenient device for opening or closing an electric current. It performs a similar service to that of a valve in a water system, except that it has no positions corresponding to partly open. It must be completely open or completely shut.

Q. What is a single-pole switch?

A. One which opens one wire of a circuit.

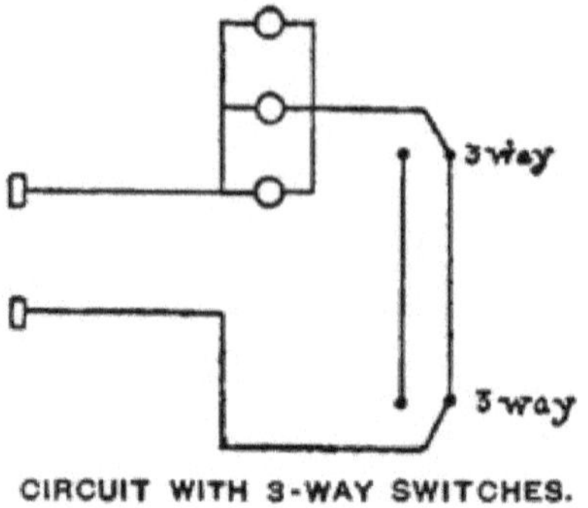

CIRCUIT WITH 3-WAY SWITCHES.

Q. What are double- and triple-pole switches?

A. Those which open two or three wires of the circuit.

Q. When are three-way switches used?

A. When it is desired to control lamps from either of two points.

Q. In calculating the carrying capacity of switches, what general rules are employed?

A. Where current goes through solid metal allow one square inch per 1000 ampères, and where it goes through the joint between two pieces allow one square inch of contact surfaces to each 75 ampères.

ELECTRIC LIGHTING.

Q. In what ways may arc lamps be classified?

A. (1) According to the kind of distribution-system for which they are intended, as constant potential arc lamps and series arc lamps; the latter are in general used now only by central stations. (2) According as they are to be supplied by direct or alternating current, into direct-current arcs and alternating arcs. (3) According to the degree of enclosure of the arc, into open arcs and closed arcs.

Q. What are the requirements of all arc lamps?

A. All lamps to be commercially satisfactory must do two things: They must strike the arc—that is, after current has commenced to flow they must automatically draw the carbons apart so as to *start* the arc. They must also regulate—that is, as the carbons burn away they must be automatically fed together, and the feeding of one must not appreciably affect the brilliancy of others.

Q. How are these accomplished in an arc lamp burning on a parallel or constant potential system of distribution?

A. The current coming from the line to the positive lamp-terminal passes through a coarse wire coil and then through a chain or brush contact to the upper carbon, through the upper and lower carbons, and back through a wire resistance, which can be varied, to the other terminal of the lamp and thence to line. The passage of current through the coil lifts an iron armature or core, as the case may be, to a certain distance depending on the strength of the current. This armature lifts a clutch-device which raises the upper carbon. The arc is thus struck and the lamp continues to burn, the two carbons being gradually consumed and the arc becoming longer. As the arc lengthens its resistance becomes greater and the current less. This allows the armature to drop down a little, and the clutch tripping against a stop lets the upper carbon slide through a little, thus shortening the arc. The moment the arc has been shortened sufficiently to increase the current enough to lift the clutch off the tripping-stop the feeding of the carbon ceases and the lamp continues to burn till the arc again becomes too long.

Q. Can two or more of these lamps be placed in series?

A. No; when several lamps are to be operated in series they will not all feed at the same time, so that the action of one would interfere with the others unless some different arrangements were introduced.

Q. What modification of the mechanism is made when lamps are to be run in series?

A. An additional magnet with fine wire coil is connected as a shunt around the arc, and its armature arranged so that when lifted to a certain point it makes the clutch feed. As the arc lengthens its resistance increases, and also the pressure between its terminals. Hence more current is sent around the fine wire coils, raising their armature and starting the feeding mechanism.

Q. What is the difference between open and closed arc lamps?

A. An open arc lamp is one in which the air has free access to the arc. A closed arc lamp is one in which a small inner globe placed around the arc prevents, to a great extent, the access of air..

Q. What is the object of enclosing the arc?

A. The consumption of carbon is diminished and the light is steadier.

Q. How long do carbons last in the two types of lamp?

A. About 7 hours in the open arc and about 100 hours in the closed arc.

Q. How are lamps rated commercially?

A. Lamps are rated in candle-power according to their brilliancy in the angle of greatest brilliancy. Thus the ordinary street lamp rated at 2000 candle-power gives that brilliancy only at an angle from the horizontal of about 45 degrees. At any other angle its brilliancy is less, and the average candle-power below the horizontal will not be much over 800 candle-power. Such a lamp requires a current of 9.6 ampères and about 45 or 50 volts, and a lamp using such current and pressure that their product is 450 watts may be considered commercially a 2000 candle-power lamp.

Q. What current does a nominal 2000 candle-power closed arc take?

A. About 5 ampères on steady burning, though nearly double this on first starting.

Q. What is the voltage between the carbons?

A. About 80 to 90 volts.

Q. What effect does the use of two globes have on the distribution of light?

A. It is more even with the closed arc on account of the two globes, but for the same reason a larger percentage of light is absorbed.

Q. What are the essential features of the incandescent lamp?

A. Incandescent lamps consist of a carbon filament attached to platinum wires, which is mounted in a glass globe from which the air has been exhausted and which is sealed up so as to exclude air. The platinum wires serve to connect the filament to the terminals of the lamp base. The vacuum is made as perfect as possible, so that there may remain no air inside the globe in which the highly heated filament would burn away.

Q. How is the filament made?

A. By taking a slender piece of some material consisting largely of carbon, such as bamboo, silk, paper, or cellulose, and heating it intensely in a furnace so as to drive out all the other material, leaving a very nearly pure carbon thread. In order to smooth out the roughness and make its section uniform at all points, a current is passed through it large enough to heat it to nearly a white heat in an atmosphere of some hydrocarbon, like coal gas. This causes carbon to be deposited most largely at the hottest points, which are those of the smallest cross-section. The filament is then attached to the platinum leading-in wires and placed in the globe.

Q. What is the remainder of the process of making the lamp?

A. A mechanical air-pump exhausts the air from the globe, and, finally, by passing a strong current through the filament, the latter, heated to incandescence, burns away the remnant of oxygen remaining. The bulb is then sealed up and the platinum wires connected to the lamp-base terminals. Finally, the lamps are tested to see at what voltage they will give the candle-power for which they are intended.

Q. What is the effect of use on the lamp?

A. Its candle-power gradually diminishes owing

to the deposition of carbon from the filament on the walls of the globe, the layer of carbon absorbing the light-rays, so that after a few hundred hours' burning the lamp must be replaced by a new one.

Q. What candle-powers are ordinarily made?

A. 8, 10, 12, 16, 20, 24, 32, 50, 100, 150, though the last two sizes are rarely used, arc lamps being employed instead.

Q. What are the voltages commonly made?

A. From 50 to 60, 70 to 80, 100 to 120, and 200 to 250 lamps of 110 and thereabouts being the most common.

Q. Why are 220-volt lamps employed?

A. To secure economy in the size of the distributing wires.

Q. Why are they not more extensively used?

A. Because they are inferior in quality to the lower voltage lamps.

Q. What are the two important qualities of an incandescent lamp?

A. Its length of life and its efficiency.

Q. What is meant by efficiency?

A. The number of watts power which must be supplied to the filament to produce 1 candle-power. The most efficient lamp is that one which produces 1 candle-power with the *least* number of watts.

Q. Is there any relation between life and efficiency?

A. Yes; a somewhat unfortunate one, since we cannot improve one without injuring the other. The efficiency increases with the temperature of the filament, while the life is correspondingly diminished.

TABLE

OF EFFICIENCIES AND LIFE OF INCANDESCENT LAMPS.

Efficiency. Watts per candle.	Life-hours.	Watts per 16 c. p. lamp.	Ampères for 16 c. p. 110-volt lamp.
2.6	400	41.8	.38
3.1	600	49 6	.45
3.6	800	57.6	.52
4.0	1000	64.0	.60

Q. When is it desirable to use a low and when a high efficiency lamp?

A. It depends upon the cost of power. If coal is cheap, it pays to use a low efficiency and long life. If coal is dear, the high efficiency lamp should be used, provided the speed regulation of the engine is good enough to prevent fluctuations in the voltage of the dynamo, it being understood that any rise in voltage above that for which the lamp is intended shortens its life very seriously. Of course, where all the exhaust steam of the generator engine is used in steam heating it is desirable to use the low efficiency and long-life lamps.

ELECTRIC MOTORS.

Q. How does a motor differ from a dynamo, as regards the purpose for which it is used?

A. A dynamo transforms mechanical energy into electrical energy. A motor transforms electrical energy into mechanical energy.

Q. How do *direct-current* motors differ from dynamos, as regards construction?

A. Practically any direct-current dynamo, if current be supplied to it, will operate as a motor, and a well-designed dynamo will make a good motor. Certain alterations in winding and in other details are made in motors to improve certain qualities that may be specially desired.

Q. Will a dynamo used as a motor run in the same direction that it had as dynamo?

A. A series dynamo, when used as a motor, will run in the opposite direction, and a shunt motor will run in the same direction.

Q. What must be done to reverse the direction in which a motor will run?

A. Change the connections so as to reverse the direction of current through either (but not both) field or armature. It may further be necessary to shift the brushes to prevent sparking.

Q. When are series motors employed?

A. The series motor is used where it is necessary to start with full load and where automatic regulation for constant speed is not necessary, a hand regulation being used, as, for example, in hoists, cranes, street railways, etc.

Q. When are shunt motors used?

A. A shunt motor is used where automatic regulation for constant speed is desired. A good shunt motor will not change its speed more than 5 per cent. when the load is varied from zero to a maximum.

Q. Under what circumstances would compound motors be desirable?

A. Compound motors are used where closer speed regulation than that given by shunt motors is desired, and in special cases, such as on planers where it is desired to check the sudden large flow of current during reversal.

Q. With a series motor, whose use is almost entirely on constant pressure circuits, how is regulation of speed accomplished?

A. There are two common methods:

1. To change the pressure supplied to it, by putting in series with the motor a rheostat in which more or less pressure is used up according to the position of the rheostat-handle. Lowering the pressure will, of course, lower the speed.

2. To change the strength of the field of the

motor. This is done by winding the field coils in sections and bringing out the ends to a sort of commutating device called a controller. In one position of the controller handle the sections will all be in series, cutting down the current and making the ampère turns of the field, and hence its strength, low. In the next position, for example, three sections will be in series and three others in series, and the two sets of three in multiple, which will diminish the resistance, let more current through, and increase the ampère turns. Another position will put more in multiple and less in series, and so on till the final step puts all the sections in multiple, giving the lowest possible resistance, highest number of ampères, greatest number of ampère turns, and strongest field. With the series motor on constant potential circuits the speed is increased in proportion as we increase the field strength. A combination of the two methods is frequently used, the resistance being used during the first positions in order to cut down the excessive flow of current on starting.

Q. How are shunt motors, on constant pressure circuits, regulated for changes in speed?

A. By putting resistance coils in series with the armature and throwing more or less of them in according as we want lower or higher speed. Another method is to put a rheostat in the field circuit and vary the current flowing around the field coils by means of it.

Q. What effect does weakening the field have on the speed of the series motor on constant pressure circuits?

A. It lowers the speed.

Q. What is the effect with a shunt machine?

A. Weakening the field increases the speed.

Q. How are compound motors regulated?

A. Generally like shunt motors; but in some special cases the series coils are wound in sections and thrown in series, and finally in multiple, as is the case with series motors.

Q. In starting shunt or compound motors what precaution is necessary?

A. It is necessary to put a considerable resistance in series with the armature, on account of its very low resistance, which will vary from $\frac{1}{100}$ to $\frac{1}{1000}$ of an ohm or less, according to its size. Such a low resistance thrown across 110 volts would cause an enormous current, which would injure the commutator and brushes by sparking and the armature coils by heating. As the machine speeds up the resistance may be cut down, because the armature, which is turning in a magnetic field, produces an electro-motive force opposite to that of the circuit, which tends to cut the current down.

Q. What further protective devices are needed with motors?

A. All motors need to be protected from the danger of being overloaded. An overload, by slowing down the motor, diminishes the back electro-motive force and therefore allows an excessive current to flow, which, if long continued, would burn out the armature. The protection formerly used was a pair of fuses, one in each of the circuit wires, which were of such a size that they were expected to blow at any current exceeding that corresponding to the maximum load for which the motor was designed. Owing to the uncertain action of fuses, a circuit-breaker is now almost universally used, mounted on the starting-box. Another thing which must be guarded against is this: Suppose that the circuit to which the motor is connected is overloaded, perhaps by some accident, and the circuit-breaker of that circuit on the switchboard should open. This would cut off current from the motor and it would stop. Now if nothing were done except at the switchboard to throw in the circuit-breaker again, we should throw the full voltage on the motor armature, none of the rheostat being in series with it, as it had been previously cut out of the circuit when the motor was first brought up to speed. The result, of course, would be a tremendous flow of current and injury to commutator, brushes, and perhaps the armature, depending upon how quickly some one opened the switch which connected the motor to the circuit. To obviate this difficulty, the rheostat arm has attached to it a spring which tends to pull it back to the position in which all of its coils are in series with the armature. At the other limit of its motion, where it would stand when all the coils had been cut out of the circuit, is a magnet wound with fine wire and supplied from the circuit wires. When the rheostat arm gets to this position the magnet holds it there by its attraction for a piece of iron mounted on the arm, as long as the current flows through the coil; but if the circuit-breaker goes off or the voltage disappears for any reason, the magnet lets go and the spring pulls the rheostat arm back to the position of safety.

Q. What are the commercial sizes in which motors are built?

A. $\frac{1}{12}$, $\frac{1}{6}$, $\frac{1}{4}$, $\frac{1}{2}$, 1, 2, 3, 5, $7\frac{1}{2}$, 10, 15, 20, 25, 50, 75, 100, and upward.

Q. What are the standard voltages?

A. 110 to 125, 220 to 250, and 500 to 550.

Q. What is a motor-generator?

A. A combination of motor and generator on the same shaft. The most easily understood form would be a motor which might be designed for any voltage, speed, and power, coupled directly to the shaft of a dynamo designed for the same speed, but for any voltage and the same output as the motor. Such a machine has two distinct commutators, brushes, armatures, and fields.

Q. How is this arrangement modified in practice?

A. By using a common armature core and field, and putting the two sets of armature windings on the same core, insulated, of course, carefully from each other.

Q. What are some of its principal uses?

A. 1. To change from a high pressure and small current to a lower pressure and correspondingly greater current.

2. With its generator armature in series with some circuit to raise the pressure of that particular circuit higher than that of the other circuits supplied from the principal generator. In such uses it is called a booster.

3. In connection with storage batteries, it being used in series with the charging mains to increase the pressure in proportion as the batteries become more fully charged.

It is also used to a considerable extent in telephone exchanges for operating the calling circuits, the generator end being arranged to give an alternating current.

STORAGE OR SECONDARY BATTERIES.

Q. Of what does the storage battery, as commercially sold, consist?

A. Of two lead plates, or sets of plates, immersed in a jar containing dilute sulphuric acid, the plates having the form of grids, the holes in which are filled with active material.

Q. Of what does this active material consist?

A. On the positive plate, of peroxide of lead. On the negative plate, of metallic lead in finely divided, spongy condition.

Q. What do you mean by the positive plate?

A. Just as with any battery, the plate *from* which current will flow through a conductor connecting it to the other plate.

Q. How can you tell by the eye which is the positive plate of a storage cell?

A. By its reddish color.

Q. Is there any other way?

A. Yes; there is always one more negative plate in a cell than there are positive plates.

Q. Are the positive and negative plates in contact?

A. The positives are all joined to each other, likewise the negatives; but the positives are separated from the negatives by about $\frac{1}{8}$ of an inch, the space between being filled with sulphuric acid.

Q. What do you mean by the discharge of a cell?

A. Allowing it to furnish current, as it will do if the positive and negative terminals are connected by a conductor.

Q. What are, roughly, the chemical changes that take place during discharge?

A. The peroxide on the positive is changed to lead sulphate. The spongy lead on the negative is likewise changed to lead sulphate.

Q. What do you mean by charging a cell?

A. Running a current from some generator through the cell in the opposite direction to that of the current which it furnished during discharge.

Q. What chemical action takes place?

A. The reverse of what occurred during discharge. On the positive plates lead sulphate is changed to lead peroxide and on the negatives to metallic lead.

Q. What pressure is furnished by such a storage cell?

A. When fully charged, about 2.2 volts. This gradually diminishes during discharge to 1.8 volts beyond which point further discharge would injure the cell.

Q. What are the principal sources of trouble, and how are they remedied?

A. The principal troubles of storage cells are *short-circuiting*, *buckling*, and *sulphating*. The first is caused by buckling of plates or by the dropping out of portions of the pencils of active material, which in time form between the positive and negative plates a connection which causes loss of charge and destruction of the plates if not noticed and remedied by taking out the material. Buckling is due to an excessive rate of discharge or an unequal discharge at different parts of the plate. To assist in preventing it the plates are separated by glass or rubber distance-pieces. Sulphating, or the production of a complex, hard, white lead sulphate, is caused by carrying the discharge of the battery too far or by letting it stand too long without recharging. It is remedied by persistent charging.

Q. What are the principal advantages of using storage cells?

A. To take care of light loads, thus permitting dynamos, engines, and perhaps a boiler to be shut down; to maintain a steady pressure; and to take care of the "peak of the load," * thus enabling the machinery to work at a more even load and securing greater economy.

* See "Roper's Engineers' Handy-Book," page 755.

Q. How are storage cells rated?

A. By their capacity in *ampère-hours.* Thus, a cell of 50 ampère-hours is one which when discharged at its normal rate gives out such a number of ampères for such a number of hours that the product of the number of ampères by the number of hours equals 50. The capacity of a cell, or the number of ampère-hours which can be taken from it without carrying the voltage lower than 1.8 volts, is very much affected by the rate of discharge, being much less at a rapid than at a slow rate of discharge.

Q. What is the efficiency of a storage cell, and how is it measured?

A. The efficiency of a cell is the ratio between the amount of power which can be taken out of it and that which is put into it. It, like capacity, varies with the rate of discharge, and may be anywhere from 50 to 95 per cent., according to the charge and discharge rates used. Eighty per cent. for the normal discharge-rate of a cell is a good value except for the very largest cells. To measure the efficiency the watt-hours put in during charge are measured by an ammeter and voltmeter, and, similarly, the watt-hours taken out in discharge. The quotient of the latter by the former is the efficiency.

METHOD OF CONNECTING STORAGE BATTERIES.

Owing to the fact that the electro-motive force of a cell increases with charge and diminishes with discharge, it is necessary to have special arrangements by which a dynamo while supplying lights may charge a battery of cells, and by which the electro-motive force of a set of cells may be kept constant while they are supplying lamps. The arrangement for discharge will be first described. Supposing a 110-volt system, we must have a number of cells in series equal to $\frac{110}{1.8}$ volts, or about 60 cells. When fully charged, as each cell has an electro-motive force of 2.2 volts, the total electro-motive force of the 60 cells would be 132 volts, a pressure which would seriously injure the lamps. When the cells are fully charged, therefore, a sufficient number are switched out of circuit to bring the pressure down to 110 volts. As the cells discharge and their electro-motive force falls, these cells are switched back into the circuit one at a time, till at the end of the discharge they are all in circuit.

In charging, the electro-motive force rises. As it is desired to run 110-volt lamps and charge the cells at the same time, we cannot raise the pressure of the lighting dynamo; so an auxiliary dynamo or booster is employed, its armature being put in series with the cells and its field varied by its rheostat so as to give enough additional volts for charging at the proper rate. The accompany-

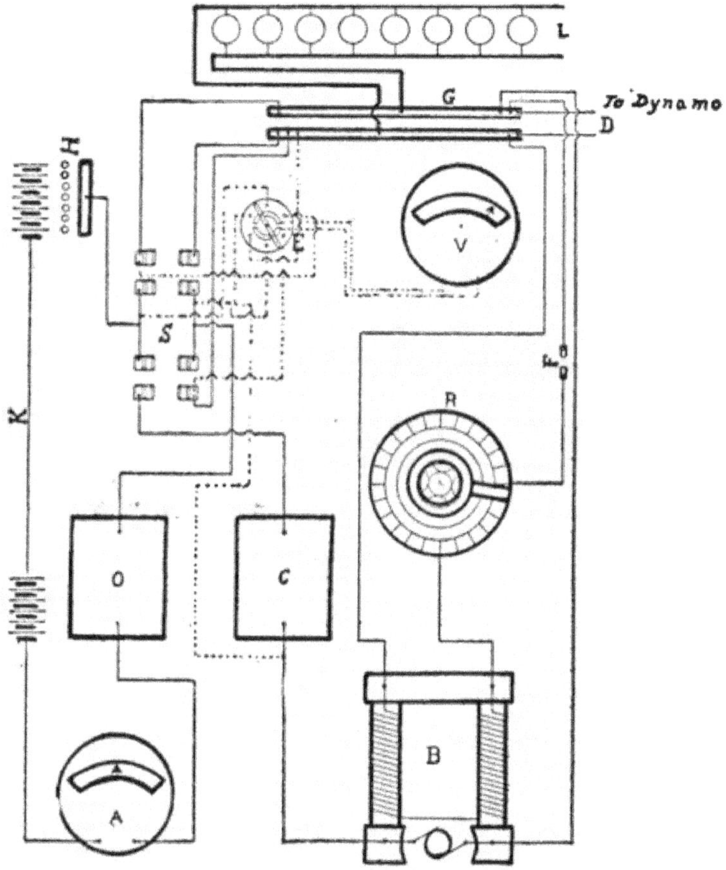

ing diagram of connections shows the arrangement. *B* is the booster and *R* its rheostat. *V* is a voltmeter and *A* an ammeter, so arranged that its needle stands in the center of the scale when no current is flowing through it, moving to one side for a charging current and to the opposite side for a discharge current. *K* represents the main battery and *H* the switch which throws the reserve cells in and out. *S* is a double-throw switch, which in one position connects the batteries to the lamp to be supplied with current, and in the other position connects it to the dynamo for charging. *E* is a switch for connecting the voltmeter, so as to give the voltage of the battery, the line, and the charging dynamo and booster respectively. *O* is an automatic circuit-breaker, which will operate if too great current is taken out of the batteries, and *C* is a circuit-breaker which will open the circuit if the charging current becomes *less* than a certain value. This last is necessary if a compound-wound dynamo is used in order to protect the dynamo from having a reverse current sent through it from the battery if by accident it was slowed down or stopped before the charging switch had been opened.

Several other arrangements are employed; but a proper understanding of the one described above will be sufficient to enable the engineer to comprehend the others without difficulty.

ELECTRIC SIGNALS.

Q. Of what four elements are most signal systems made up?

A. Of the battery, line, the operating station, and the receiving mechanism.

Q. What is the function of each element?

A. The battery furnishes the electrical energy for operating the signals, and the line serves to transmit this energy. The operating station, which generally consists of a key, a switch, or a push-button, closes the electrical circuit and permits the operating current to flow. The receiving stations are somewhat varied in design. They may consist of a bell or telegraph sounder, giving the signals by sound, or of a galvanometer or a shutter-drop, which conveys the signals by means of sight. Frequently the two methods of sound and sight are combined.

Q. Of what does an electric bell consist?

A. Of an electro-magnet, to the armature of which is connected a hammer arranged to strike a gong when the armature is pulled up to the core of the magnet by the passage of an electric current. When current ceases the magnet loses its strength and a spring pulls the armature away from the core and also the hammer from the gong.

Q. Into what classes are bells divided?

A. Into *single-stroke* bells, which make but one stroke each time that circuit is closed, and *vibrating bells*, whose hammer continues to vibrate as long as circuit is closed.

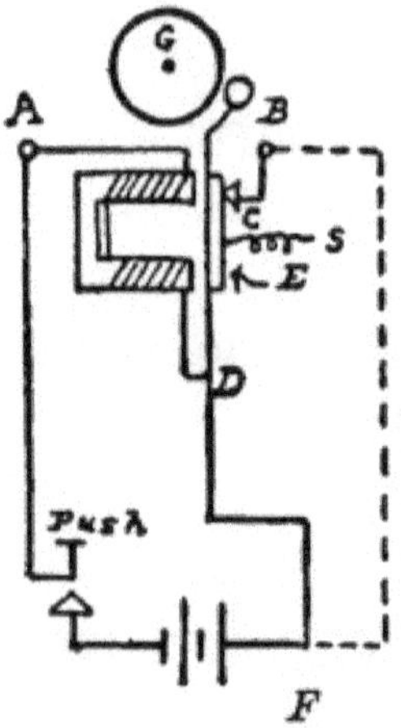

Q. How is a single-stroke bell connected?

A. As shown by the solid lines in the cut.

Q. How is a vibrating bell connected?

A. As shown in the cut, the connection *F–D* being considered as removed.

Q. Explain the complete action of the vibrating bell.

A. When the button is pressed down, the circuit being closed, current will flow from *F* to *B*, *B* to the contact point *C*, through the armature *E* to *D*, from *D* through the magnet coil to *A*, and from *A* back through the closed push and battery to *F*. Owing to the current, the electro-magnet pulls the armature *E* toward itself and the hammer strikes the gong *G*; but as soon as the armature moves toward the magnet the circuit is opened, because *C* no longer touches *E*. The current therefore stops, and as the electro-magnet no longer has any strength the armature is pulled away from it by the spring *S*. This movement, however, brings *E* and *C* into contact again, causing the whole action to be repeated, and this continues as long as the push-button is held down, provided the battery keeps up its strength.

Q. What three styles of bells are there?

A. *Wooden box*, the working parts of which are covered with wood; *iron box*, when they are covered with iron, and *skeleton frame*, when they are not covered at all.

Q. Show how you would connect three bells to ring by one push-button.

A.

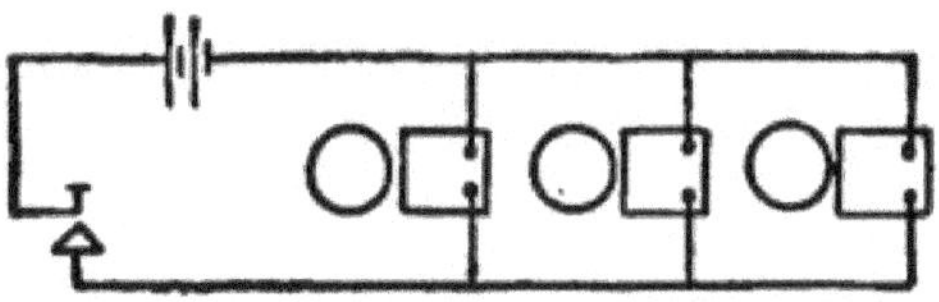

Q. Show how to connect two bells to be rung by either of two pushes.

A.

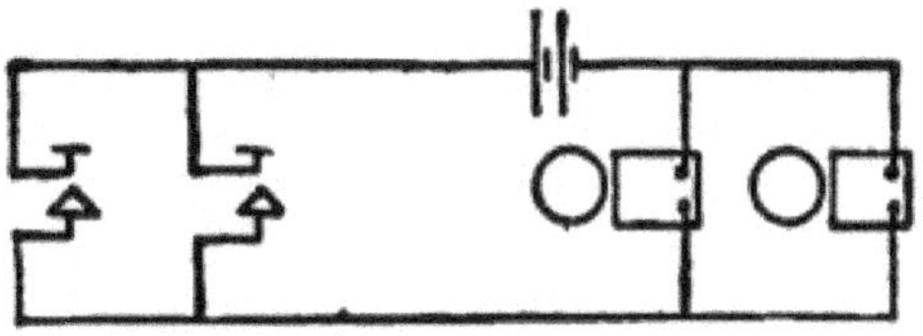

Q. Show how you would connect a *return call* between two points.

A.

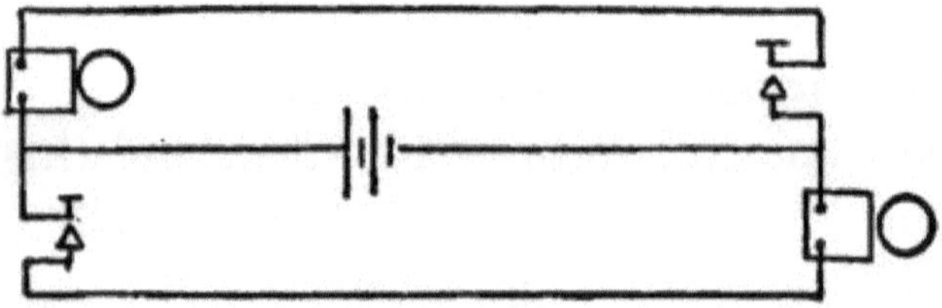

Q. What is an annunciator?

A. The annunciator in principle consists of a number of bells mounted together in a case, each operated by its own push located in some distant place. In practice, however, it would be difficult to tell from the sound of the bells which station was calling, so the hammers and gongs are omitted, and instead we have a simple mechanism operated by the armature, called the drop.

Q. Explain the details of one form of drop.

A. It consists of a coil whose armature is an iron rod which is sucked up into the coil when current passes through it. This releases a pivoted needle, which is hung eccentrically so that it turns from the horizontal to the vertical position. Each needle being numbered or otherwise marked the point from which the signal was sent is, of course, known.

Q. How are the needles restored?

A. By a rod carrying little stops, which when pushed up force the needles back to their original position.

Q. What is an automatic set-back annunciator?

A. One in which this rod is lifted by an electro-magnet so connected that current flows through it when any push-button is pressed. All the needles are pushed back to their horizontal position, after which the needle corresponding to the push-button last pressed turns to the vertical position.

Q. Show by a diagram the connections for an automatic set-back annunciator system.

A.

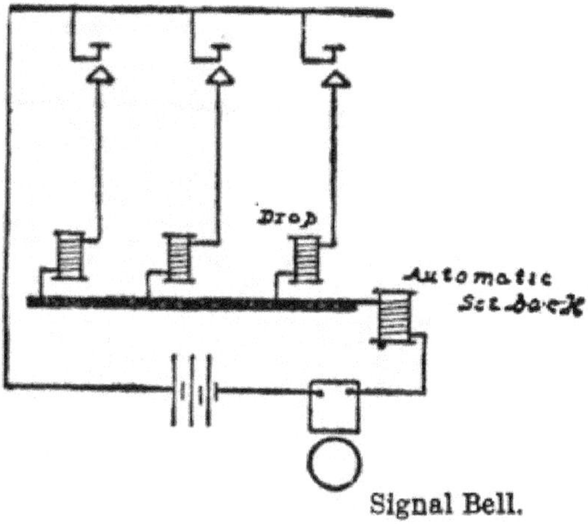

Q. How does the return-call annunciator system differ from this?

A. By the addition of another wire between each push-button and the annunciator.

Q. What is a fire-alarm attachment?

A. A device, frequently added to annunciators for use in hotels, which closes the circuit of the bells in the rooms, the effect being the same as if all the return-call pushes on the instrument were pressed simultaneously.

Q. How does a burglar-alarm system differ from the ordinary annunciator system?

A. Burglar-alarm systems are similar to simple annunciator systems, with the addition of a bell in an auxiliary circuit which is closed when any of the drops operate. This auxiliary bell will therefore continue to ring till some one comes along and restores the drops to their usual position with the needles horizontal. The push-buttons are of a somewhat modified pattern and are placed in doors and window-casings, so that if either a door or window is opened the contacts of the button touch each other and close the circuit, causing the corresponding drop on the instrument to operate. Frequently the pushes of all the windows and outside doors of any one room are connected in multiple on one circuit, so that any one of them when closed operates the drop corresponding, it not being necessary to have a drop for each window and door, but only for each room.

Q. Why are watchmen's clock systems used?

A. To insure that watchmen make their rounds at the time and in the order that they are expected to do so.

Q. Into what classes may they be divided?

A. Into the *battery* and *magneto* systems, according as the energy for actuating the recording device is obtained from a battery or from a small dynamo.

Q. Explain the arrangement and operation of a battery system.

A. This system is wired like a simple annunciator system. Its push-buttons are of such pattern that circuit will be closed in them only by pushing into them a special key carried by the watchman. The annunciator of the ordinary system, with slight modification, becomes the watchman's clock, the signal bell and self-restoring magnet of the annunciator being omitted. The armature of each drop is made to actuate a little needle which punctures a hole in a paper recording dial. This dial being divided in spaces corresponding to the hours from 12 o'clock to 12 o'clock, and being further subdivided into spaces corresponding to five minutes, and rotating so as to make one complete turn in the 12 hours, the position of the punctured holes on the paper tells at what time they were made by the watchman. The dial has also a number of circles marked on it corresponding to the number of stations, and each needle pricks its holes in one of the circular spaces formed by these rings, so that a hole in a certain ring means that the key has been put in the corresponding station push-button.

Q. What is the weak point of this system?

A. That if the watchman can get at the two wires leading to any station and can connect them together, he can make the clock register as if he had actually gone to that station.

Q. How does the magneto system differ from it?

A. The wiring and clock are the same; but instead of the special push-button to be operated by a key, a little dynamo, called a magneto, is placed at each station. The watchman carries a handle which he puts on a stud connected with the shaft of the dynamo armature. Turning the handle sends a current through the coil corresponding at the clock and causes the needle to make a record.

Q. What are the advantages of the magneto system?

A. There are no batteries to be taken care of and the watchman practically cannot make a proper record without going to the station.

Q. What kind of batteries are used for operating the above systems?

A. Some form of the zinc-carbon sal-ammoniac cell.

Q. How many are required for the different systems?

A. For single bells or annunciators with short circuits, as in a dwelling-house, three cells are usually sufficient. For larger buildings five or six will be needed. For automatic fire-alarms a much larger number is needed, the exact number being stated by the manufacturer, as a rule. For burglar-alarm and watch-clock systems six are, as a rule, sufficient, and sometimes a less number may be used.

THE TELEPHONE.

The phenomenon of sound is caused by vibrations of the particles of air; its pitch is dependent upon the *number* of vibrations per second, its loudness on the *wideness* of those vibrations, and its quality, that property by which we distinguish tones of the same pitch and loudness, upon the *form* of the vibrations. This last point is some-

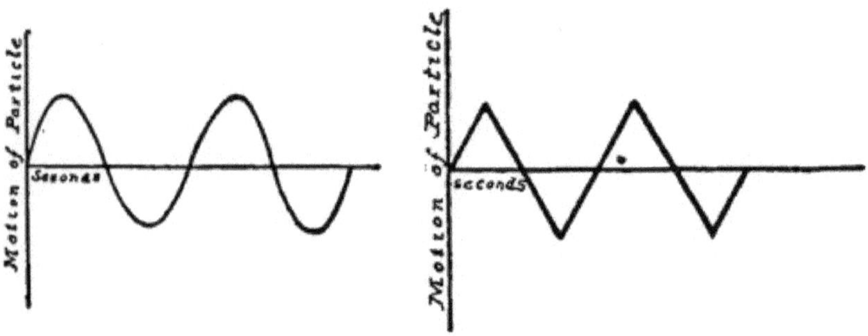

what difficult to understand. Suppose that a mass of air is set in vibration by a tuning-fork, and that we study the motion of a single particle of air by plotting on a flat surface. Let distances to the right of the vertical represent time, and vertical distances represent the distance which the particle has moved through at any time. The motion of the particle would be represented by the wavy line in the figure. Distances above the horizontal correspond to motion in one direction from its position of rest, and distances below the horizontal represent, similarly, motion in the opposite direction. If we set the air into vibration by means of a bowed violin-string, the shape of the wavy line would be very much altered, as in the second figure. To perfectly reproduce sounds it is necessary to reproduce the pitch or number of waves per second and the quality or form of these waves, and sufficient wideness of vibration to affect the listening ear.

The telephonic transmission of speech between two points may be best considered in two parts: (1) The transmitter, which produces in the wires connecting the two points a varying current whose curve of variation, if plotted, has the same number of vibrations per second, and whose form is the same as that of the sound-waves which strike upon the diaphragm of the transmitter mouthpiece. (2) The receiver, into which comes this varying current, which is made to set a diaphragm into vibrations exactly similar to those of the transmitter diaphragm. The receiver diaphragm, of course, sets the air surrounding it into vibrations similar to those caused by the voice speaking, and the ear of the listener is affected in the same way, though not so strongly as if the speaker were talking directly to him.

Q. Describe the magneto receiver.

A. The magneto receiver consists of a bar magnet with a short cylindrical pole-piece of soft iron on one end. Mounted on this pole-piece as an axis is a little wooden spool wound with fine wire. In front of the spool is a thin circular disk of soft iron.

Q. What improvements have been made in the receiver?

A. It is now made with a magnet of horse-shoe pattern, each pole having a spool of wire on it.

Q. What was the original form of the transmitter?

A. Originally the same instrument was used alternately as transmitter and receiver.

Q. Explain the operation when two of these receivers are connected together by two wires, one being spoken into and the other serving as a receiver.

A. The voice of the speaker sets the diaphragm of the transmitter into vibration. The motion of the iron near the magnet-pole alters the position and density of the magnetic lines of force enclosed by the coil and sets up a varying electro-motive force in the coil. This produces a current in the line with a variation or wave-form similar to the original sound-wave. This varying current flowing around the coil of the *receiver* causes the strength of its pull on the receiver diaphragm to vary in a similar way, and therefore to set up in the receiver diaphragm vibrations similar to those of the transmitter diaphragm. This sets the surrounding air into similar vibration. This causes the listener's ear to be affected just as if the speaker were talking directly in his ear, although not so loudly.

Q. What form of transmitter is now used?

A. That which is known as the battery or carbon transmitter.

Q. Explain how it differs from the magneto transmitter.

A. In the magneto transmitter just described the varying current is produced by setting up an electro-motive force whose wave-form of variation is similar to that of the sound-wave producing

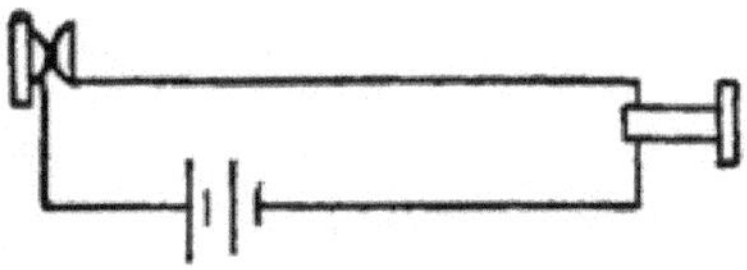

CARBON TRANSMITTER AND CIRCUIT.

it. Another way to produce the varying current is to use a constant electro-motive force, but employing a resistance varied by the sound-wave and having the same wave-form of variation. A current is sent through the circuit consisting of the receiver, line, and carbon contact, as shown in the diagram. One of the carbon pieces is fixed and the other moves with the diaphragm. When the latter is spoken against, its vibrations cause the varying pressures on the contact between the two carbon pieces. This causes the varying resistance, which produces the varying current necessary to transmit speech.

Q. Do the present forms of transmitter consist of a single carbon contact?

A. No; in order to make the variation of resistance as great as possible the number of contacts is increased by having the circuit pass through a number of small carbon particles against which the diaphragm presses.

Q. What is the induction coil, and why is it used?

A. On long lines the resistance of the lines, which is fixed in value, is so much greater than that of the variable carbon contacts that the effect of the latter in varying the total resistance in circuit is practically zero. To overcome this difficulty the induction-coil is used. It consists of a bundle of fine iron wires about three inches long, and wound around these as an axis is a coarse wire coil of about No. 16 wire and a fine wire coil of No. 24 or smaller, according to the length of line.

Q. How is the coil connected?

A. As shown in the diagram.

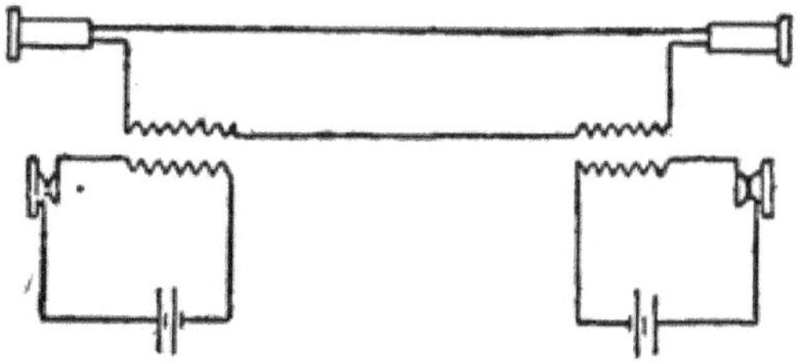

CONNECTIONS USING INDUCTION COILS.

Q. What are the methods used in calling up?

A. By a battery and ordinary vibrating bell, called the *battery call*, and by a magneto and special bell, called the *magneto call.*

Q. When is the former used?

A. Generally for distances not exceeding a few hundred feet.

Q. On what two systems are telephones operated?

A. On the intercommunicating system and on the exchange system.

Q. What is the intercommunicating system?

A. The intercommunicating system consists of instruments as above described, combined with a suitable number of wires running to all instruments, and at each instrument such a form of mechanical-contact changing switch as to enable each telephone station to call up any particular station without interfering with any others who may be talking.

Q. What is the general scheme of wiring for this system?

A. To each instrument as many wires are run as there are telephones in the system, plus two (three in some systems). These wires are preferably of different colors, to facilitate making proper connection.

Q. What kind of a call is used?

A. Either may be employed, but the battery call is more common.

Q. What requirement must a successful intercommunicating system fulfil?

A. That no other act is necessary after finishing conversation than to hang up the receiver on the hook. Some systems require that a lever shall be returned to a certain point or that a plug shall be put in a certain hole in addition to hanging up the receiver. Such systems are faulty.

Q. How many instruments are used on such systems?

A. Any number may be used, but it is rarely advisable to go above twelve or fifteen, the exchange system being preferable when a greater number is required.

Q. What is the general nature of exchange systems?

A. In such systems two (or sometimes three) wires run from each telephone to a central point, at which an operator sits, whose duty it is to connect the lines of any two telephones by means of a convenient switchboard and to disconnect them when they have finished talking. The connections are made through a pair of flexible cords, called talking-cords, which are attached to plug-shaped pieces.

Q. How are the subscribers called up?

A. By either battery or magneto call.

Q. What is the general method of operation in an exchange system when one party wishes to talk to another?

A. See "Roper's Engineers' Handy-Book," pages 771–773.

Q. May any number of instruments be connected on an exchange system?

A. Yes; the switchboard is increased as fast as the addition of instruments renders it necessary.

INDEX.

www.ingramcontent.com/pod-product-compliance
Lightning Source LLC
Chambersburg PA
CBHW021356210326
41599CB00011B/905
9783743467019